Spectral Theory of Nonautonomous Dynamical Systems and Applications

Thai Son Doan

Spectral Theory of Nonautonomous Dynamical Systems and Applications

Springer

Thai Son Doan
Institute of Mathematics
Vietnam Academy of Science and
Technology
Hanoi, Vietnam

ISBN 978-981-97-5522-6 ISBN 978-981-97-5520-2 (eBook)
https://doi.org/10.1007/978-981-97-5520-2

This Springer imprint is published by the registered company Springer Nature Singapore Pte Ltd.
The registered company address is: 152 Beach Road, #21-01/04 Gateway East, Singapore 189721, Singapore

Preface

This book is an extension of author's Habilitation thesis defended in 2016 at Technical University Dresden, Germany. The main scientific content of the book is taken from the joint work of the author with Nguyen Dinh Cong, Peter Kloeden, Peter De Maesschalck, Ken Palmer, Jeroen Lamb, Martin Rasmussen, and Stefan Siegmund. The author would like to thank them for the long-standing pleasure and inspiring scientific cooperation on the topic of nonautonomous dynamical systems.

Hanoi, Vietnam | Thai Son Doan

Introduction

The theory of nonautonomous dynamical systems has experienced a renewed and steadily growing interest in the last 20 years, stimulated also by synergetic effects of disciplines which have developed relatively independent for some time such as *topological skew product* (see, e.g., [109]), *random dynamical systems* (see, e.g., [6]), *finite-time dynamics* (see, e.g., [68]), and *control systems* (see, e.g., [41]).

The importance of nonautonomous dynamical systems is illustrated by the fact that the technological and economic development of our society has generated the need to deal with very complex systems that require an accurate level of understanding (see monograph [KR11]). The crisis of the financial markets and weather phenomena associated to climate change are examples of dynamical processes with a deep economic impact that require sophisticated models to take nonautonomous influences into account.

The main challenge in the study of nonautonomous phenomena is to understand the often very complicated dynamical behavior both as a scientific and mathematical problem. The aim in this book is to present an effort in introducing spectral theory of nonautonomous differential equations with applications to linearization of nonautonomous differential equations and spectral theory of random dynamical systems with applications to pitchfork and Hopf bifurcation with additive noise.

Spectral Theory of Nonautonomous Differential Equations

The stability theory for linear nonautonomous differential equations has its origin in A.M. Lyapunov's celebrated PhD Thesis [89], where he introduced characteristic numbers, so-called *Lyapunov exponents*, which are given by accumulation points of exponential growth rates of individual solutions. It is well-known that in case of negative Lyapunov exponents, the stability of nonlinearly perturbed systems is not guaranteed without an additional regularity condition.

In the 1970s, R.S. Sacker and G.R. Sell developed the Sacker–Sell spectral theory for nonautonomous differential equations, see [108]. In contrast to the Lyapunov spectrum, the Sacker–Sell spectrum is not a solution-based spectral theory but rather is based on the concept of an exponential dichotomy, which concerns uniform growth behavior in subspaces and extends the idea of hyperbolicity to explicitly time-dependent systems. If the Sacker–Sell spectrum lies left of zero, then the uniform exponential stability of nonlinearly perturbed systems is guaranteed.

It was shown in [90] that the regularity condition on Lyapunov exponents can be more robustly replaced by a nonuniform exponential dichotomy. Here the nonuniformity refers to time, and in contrast to that, so-called Bohl exponents, introduced by P. Bohl in 1914 [28], measure exponential growth along solutions uniformly in time. Bohl exponents have been studied extensively in the literature [54], and current research focuses on applications to differential-algebraic equations and control theory [3, 22, 72, 88, 124], and parabolic partial differential equations [91].

In Sect. 1.2, we develop the Bohl spectrum as the union of all possible Bohl exponents of a nonautonomous linear differential equation on a half line. We show that the Bohl spectrum lies between the Lyapunov and the Sacker–Sell spectrum and that the Bohl spectrum is given by the union of finitely many (not necessarily closed) intervals. Each Bohl spectral interval is associated with a linear subspace, leading to a filtration of subspaces which is finer than the filtration obtained by the Sacker–Sell spectrum. In the remaining part of this section, we analyze in detail situations in which the Bohl spectrum is identical to the Sacker–Sell spectrum, and in particular, we obtain this for diagonalizable systems, integrally separated systems, and systems with Sacker–Sell point spectrum.

Linearization Theory of Nonautonomous Differential Equations

Normal form theory dates back to Poincaré who proved in his thesis in 1879 (see, e.g., [114] and the references therein) if the eigenvalues $\lambda_1, \dots, \lambda_n$ of the linearization

$$\dot{x} = Df(x_0)x \tag{1}$$

at a rest point x_0 of an autonomous differential equation

$$\dot{x} = f(x) \tag{2}$$

satisfy the *nonresonance condition*

$$\lambda_j \neq \sum_{i=1}^{n} \ell_i \lambda_i \tag{3}$$

for $j \in \{1, \dots, n\}$, $\ell_i \in \{0, 1, \dots\}$, $\sum_{i=1}^{n} \ell_i \geq 2$, then the differential equation can be formally linearized, i.e., using a formal power series transformation. Step by step each Taylor term

$$\left(0, \dots, 0, \tfrac{1}{\ell_1! \cdots \ell_n!} D_{x_1}^{\ell_1} \cdots D_{x_n}^{\ell_n} f_j(x_0) \cdot x_1^{\ell_1} \cdots x_n^{\ell_n}, 0, \dots, 0\right)^{\mathrm{T}}$$

is eliminated. In case condition (3) is not satisfied for some j and multi-index $(\ell_1, \dots, \ell_n)$, then the corresponding Taylor term is called *resonant*, leading to a so-called Poincaré Dulac type normal form which still contains resonant terms. An extension of Poincaré Dulac type normal form results to nonautonomous differential equations based on the dichotomy spectrum of their linear part can be found in [114]. Wu and Li [126] generalize those results to analytic equivalence theorems for analytic nonautonomous differential equations in the Poincaré domain.

In the case that the system is hyperbolic and there is no resonant terms up to some order, Sternberg proved that system (2) and (1) are smoothly equivalent, i.e., system (2) can be smoothly linearized, see [116, 117]. Our aim in the first part of Chap. 2 is to extend this result to nonautonomous differential equations. More precisely, we show that a hyperbolic nonautonomous differential equation can be smoothly linearized if the associated Sacker–Sell spectrum satisfies a non-resonance condition. Some works related to the smooth linearization are: Hartman-Grobman theorems (C^0-linearization) for nonautonomous differential equations in [97], smooth conjugacy theorems of Sternberg type for almost periodic differential systems in [86], and analytic normal form results for random diffeomorphisms and differential equations in [87] and [85].

If, for example, the differential equation has a center manifold, i.e., $\lambda_j = 0$ for some j, then the linearization is not hyperbolic and the terms $(0, \dots, 0, \frac{1}{\ell_j!} D_{x_j}^{\ell_j} f_j(x_0) \cdot x_j^{\ell_j}, 0, \dots, 0)^{\mathrm{T}}$, $\ell_j \geq 2$, are resonant and the differential equation can in general only be partially linearized, see [31, 118]. So, in the second part of Chap. 2 we focus on partial linearization of nonautonomous differential equations in the plane with one center-like direction and one hyperbolic direction. A typical situation which we address—although our results are more general—is the following system

$$\dot{x} = \begin{pmatrix} a_1(t) & 0 \\ 0 & a_2(t) \end{pmatrix} x + r(t, x) \tag{4}$$

with $r(t, 0) = 0$, $D_x r(t, 0) = 0$, which we show to be C^k equivalent to the partially linearized system

$$\dot{x} = \begin{pmatrix} a_1(t) & 0 \\ 0 & a_2(t) \end{pmatrix} x + \begin{pmatrix} r_1(t, x_1, 0) \\ \frac{\partial r_2}{\partial x_2}(t, x_1, 0) x_2 \end{pmatrix}. \tag{5}$$

The spectral intervals λ_i correspond to the scalar equations $\dot{x}_i = a_i(t)x_i$ and satisfy $\lambda_2 < \lambda_1$ and certain gap conditions. If $0 \in \lambda_1$, then we have a one-dimensional center-like direction and λ_2 corresponds to a one-dimensional hyperbolic direction.

To gain smooth conjugacy between (4) and (5), we construct several near-identity transformations to simplify the planar nonautonomous differential equation step by step. For example, a transformation of the form $(t, x_1, x_2) \mapsto \begin{pmatrix} x_1 + h_t(x_1)x_2^\ell \\ x_2 \end{pmatrix}$ eliminates a term in the x_1-equation and similarly for terms in the x_2-equation. A main building block are mappings of the form

$$h_t(x_1) = \lim_{\tau \to \infty} \lim_{x_2 \to 0} \frac{\varphi_1(t, \tau) \circ \Phi(\tau, t)(x_1, x_2) - x_1}{x_2^\ell}$$

where φ and Φ are the solution operators of the old and new differential equation, respectively. We learned this technique from [95, p. 29] who describes that the limit for $t \to \infty$ is inspired by the Møller wave operator from quantum mechanics. It was also used in [30, 31] to construct smooth transformations which respect constraints or additional properties and in [33] for normal forms in singular perturbation problems.

Spectral Theory of Random Dynamical Systems

The fundamental results on Lyapunov exponents for random dynamical systems on finite dimensional systems were first obtained by Oseledets [96], which is now called the Oseledets Multiplicative Ergodic Theorem. Originally formulated for products of random matrices, it has been reformulated and reproved several times during the past 30 years.

When we apply Multiplicative Ergodic Theorem to bilinear stochastic differential equations, it is more convenient to work with the induced solutions on the projective space. The advantage is that by using the strong law of large number for martingales we obtain a Furtensberg-Khamsminskii formula to present the Lyapunov exponents of solutions in terms of the stationary distribution, see [6]. Being able to compute the stationary distribution of the associated flow of planar bilinear stochastic differential equations on the projective space, we archive explicit formulas for the Lyapunov exponents of planar bilinear stochastic differential equations in Sect. 3.3.

Although spectral studies of random dynamical systems have focused mainly on Lyapunov exponents [6], in Sect. 3.2.2 we develop an alternative spectral theory based on exponential dichotomies that is related to the Sacker–Sell (or dichotomy) spectrum for nonautonomous differential equations. This spectral theory is demonstrated to use in the characterization of pitchfork and Hopf bifurcations with additive noise in Chap. 5.

Genericity of Lyapunov Spectrum of Random Dynamical Systems

The study of the Lyapunov spectrum of linear cocycles is one of the central tasks of the theory of random dynamical systems, see [6]. In various situations it is of theoretical and practical importance to know when the Lyapunov spectrum is simple and the Oseledets splitting is exponentially separated. Recently, [5], [25], and [43] have derived some new results on genericity of hyperbolicity of several classes of dynamical systems including smooth dynamical systems and linear cocycles.

Our aim in Sect. 4.2 is to investigate the openness and denseness of the set of integrally separated systems in the space of bounded linear random dynamical systems. Concerning the discrete-time bounded linear random dynamical systems, this result was proved in [43]. An analogous result for bounded random differential equations equipped with the L^∞-metric is established in Sect. 4.2.2. We show that in the general case, the set of integrally separated systems is open and dense. An exception is the case when the base space is isomorphic to the ergodic rotation flow of the unit circle, in which the set of integrally separated systems is open but not generic.

After establishing the genericity of integral separation (uniform hyperbolicity) of bounded linear random dynamical systems, it is natural to ask about the genericity of stability of Lyapunov exponents under a small perturbation. Now, we would like to mention some publications that are related to this question.

In [107], Ruelle showed that the characteristic exponents of a random matrix product over a compact base space endowed with a probability measure depends real analytically on the data of the problem. This result was generalized in [60] to product of random linear operators having an invariant cone over a measurable base space. Concerning random dynamical systems without assuming to share an invariant cone, in general Lyapunov exponents do not depend continuously on the coefficients, see [84]. Using a similar technique as in [84], Bochi gave a characterization of discontinuity for Lyapunov exponents of SL(2; $\mathbb{R}$)-cocycles in a unpublished preprint [24]. Roughly speaking, it was showed that if a SL(2; $\mathbb{R}$)–cocycle has a positive Lyapunov exponent and has not a dominated splitting (integral separation), then the upper Lyapunov exponent function is not continuous at this cocycle.

Although Lyapunov exponents are in general not continuous, it was proved in [43] that continuity of Lyapunov exponents is a generic property on the space of bounded linear cocycles. Thereafter, it is natural to investigate higher regularity property than continuity property for Lyapunov exponents of linear random dynamical systems such as Hölder continuity, smoothness, and analyticity.

In Sect. 4.3, we show that analyticity of Lyapunov exponents is still a generic property on the space of bounded linear random dynamical systems. The main ingredient in the proof of this fact is to construct a suitable cone being invariant under a long enough iteration of an integrally separated linear random dynamical system. Note that generic bounded linear random dynamical systems are integrally separated, see Sect. 4.2. This fact together with results in [60, 107] deduces the conclusion that generically, the top Lyapunov exponent of random dynamical systems depends analytically on the coefficients. To derive the same conclusion for other Lyapunov exponents, we work with the generated random dynamical systems on the exterior power space.

Bifurcation Theory of Random Dynamical Systems

Despite its importance for applications, relatively little progress has been made toward the development of a bifurcation theory for random dynamical systems. Main contributions have been made by Ludwig Arnold and co-workers [6], distinguishing between *phenomenological* (P-) and a *dynamical* (D-) bifurcations. P-bifurcations refer to qualitative changes in the profile of stationary probability densities [93]. This concept carries substantial drawbacks such as providing reference only to static properties, and not being independent of the choice of coordinates. D-bifurcations refer to the bifurcation of a new invariant measure from a given invariant reference measure, in the sense of weak convergence, and are associated with a qualitative change in the Lyapunov spectrum. They have been studied mainly in the case of multiplicative noise [21, 52, 123], and numerically [11, 80].

In Chap. 5, we contribute to the bifurcation theory of random dynamical systems of pitchfork bifurcation with additive noise

$$\mathrm{d}x = \big(\alpha x - x^3\big)\mathrm{d}t + \sigma \mathrm{d}W_t\,,$$

and Hopf bifurcation with additive noise

$$\begin{aligned}\mathrm{d}x &= (\alpha x - \beta y - (ax - by)(x^2 + y^2))\,\mathrm{d}t + \sigma\,\mathrm{d}W_t^1\,,\\ \mathrm{d}y &= (\alpha y + \beta x - (bx + ay)(x^2 + y^2))\,\mathrm{d}t + \sigma\,\mathrm{d}W_t^2\,.\end{aligned}$$

Instead of looking at the absence of D-bifurcation of the global random attractor and P-bifurcation of stationary distribution (see [6, 50]), the bifurcation is characterized by a loss of uniform attractivity of the global random attractor, a change of experimental observability of the Lyapunov exponent, a change of the dichotomy spectrum of the linearization along the global random attractor.

Contents

List of Symbols

$\mathbb{N}_0$	The set of nonnegative integer numbers
$\mathbb{R}$	The set of real numbers
$\mathbb{R}_0^+$	The set of nonnegative real numbers
$\mathrm{GL}(d)$	The set of invertible matrices in $\mathbb{R}^{d\times d}$
$\mathcal{S}^1$	The unit circle on the complex plane
$\ln x$	The natural logarithm of a positive number x
$\ln^+ x$	The maximum between $\ln x$ and 0
$\|\cdot\|_2$	The Euclidean norm on $\mathbb{R}^d$
$\mathcal{L}(\mathbf{X})$	The set of bounded linear operators from a Banach space $\mathbf{X}$ into itself
id_X	The identical operator from X into itself
	A probability space
$\mathcal{L}^1(\Omega, \mathcal{F}, \mathbb{P})$	The set of integrable real valued functions
$\mathcal{L}^\infty(\Omega, \mathbf{X})$	The set of essentially bounded measurable functions from Ω to a Banach space $\mathbf{X}$
$\mathcal{B}(X)$	The Borel σ − algebra of a topological space X
ess sup	Essential supremum
ess inf	Essential infimum

Chapter 1
Spectral Theory of Nonautonomous Differential Equations

Abstract The central aim in this chapter is to develop a spectral theory for linear nonautonomous differential equations of the form

$$\dot{x}(t) = A(t)x.$$

In Sect. 1.1, we recall two well-developed spectral theories, namely, Lyapunov spectrum and Sacker-Sell spectra, for nonautonomous differential equations on the half real line ($I = \mathbb{R}_0^+$) and on the whole real line ($I = \mathbb{R}$). Section 1.2 is devoted to developing a Bohl spectral theory for nonautonomous differential equations. The content of this section is a part of Doan et al. (J Dyn Differ Equ 29(4):1459–1485, 2017).

1.1 Lyapunov Spectrum and Sacker-Sell Spectra for Nonautonomous Differential Equations

Consider a linear nonautonomous differential equation on $\mathbb{J}$, where either $\mathbb{J} = \mathbb{R}_0^+$ (the half real line) or $\mathbb{J} = \mathbb{R}$ (the whole real line), of the form

$$\dot{x}(t) = A(t)x \,, \tag{1.1}$$

where $A : \mathbb{J} \to \mathbb{R}^{d\times d}$ is a locally integrable matrix-valued function, i.e., for any $a, b \in \mathbb{J}$ with $a < b$, we have $\int_a^b \|A(t)\| \, \mathrm{d}t < \infty$. Let $X : \mathbb{J} \to \mathbb{R}^{d\times d}$ be the *fundamental matrix* of (1.1), i.e., $X(\cdot)\xi$ solves (1.1) with the initial value condition $x(0) = \xi$, where $\xi \in \mathbb{R}^d$. Denote by $\Phi(\cdot,\cdot) : \mathbb{J} \times \mathbb{J} \to \mathbb{R}^{d\times d}$ the *evolution operator* of (1.1), i.e.,

$$\Phi(t,s) = X(t)X(s)^{-1} \qquad \text{for all } t, s \in \mathbb{J}.$$

T. S. Doan, *Spectral Theory of Nonautonomous Dynamical Systems and Applications*, https://doi.org/10.1007/978-981-97-5520-2_1

1.1.1 Lyapunov Spectral Theory

Lyapunov exponents have been introduced in 1892 by Lyapunov in his doctoral thesis [89], which lead to a corresponding Lyapunov spectral theory. Roughly speaking, Lyapunov exponent of an arbitrary solution measures the asymptotic growth of the norm of this solution. Precisely, we have the following definition.

Definition 1.1 (Lyapunov Spectrum) The *lower* and *upper characteristic Lyapunov exponents* of a particular nonzero solution $X(\cdot)\xi$ of (1.1) are defined by

$$\chi_-(\xi) := \liminf_{t\to\infty} \frac{1}{t} \ln \|X(t)\xi\|$$

and

$$\chi_+(\xi) := \limsup_{t\to\infty} \frac{1}{t} \ln \|X(t)\xi\|\,.$$

The *Lyapunov spectrum* of (1.1) is then defined as

$$\Sigma_{\mathrm{Lya}} := \bigcup_{\xi\in\mathbb{R}^d\setminus\{0\}} \{\chi_+(\xi)\}\,.$$

It is well-known [1, 18] that there exist $n \in \{1, \dots, d\}$ and $\xi_1, \dots, \xi_n \in \mathbb{R}^d \setminus \{0\}$ such that

$$\Sigma_{\mathrm{Lya}} = \bigcup_{i=1}^{n} \{\chi_+(\xi_i)\}\,.$$

1.1.2 Sacker-Sell Spectral Theory

In contrast to the Lyapunov spectrum, the Sacker–Sell spectrum is based on a hyperbolicity concept for nonautonomous differential equations, given by an exponential dichotomy. In what follows, we recall the notion of exponential dichotomy; see also [47]. For this purpose, we first introduce the notion of an invariant projector associated with (1.1). An invariant projector of (1.1) is defined to be a function $P : \mathbb{J} \to \mathbb{R}^{d\times d}$ of projections $P(t)$, $t \in \mathbb{J}$, such that

$$P(t)\Phi(t,s) = \Phi(t,s)P(s) \quad \text{for } t, s \in \mathbb{J}.$$

Note that P is continuous due to the identity $P(t) = \Phi(t,s)P(s)\Phi(s,t)$.

Definition 1.2 (Exponential Dichotomy) The linear differential equation (1.1) admits an *exponential dichotomy* with growth rate $\gamma \in \mathbb{R}$ if there exist an invariant projector $P \in \mathbb{R}^{d\times d}$ and constants $K \geq 1$ and $\alpha > 0$, such that for $t, s \in \mathbb{J}$

$$\|\Phi(t,s)P(s)\| \leq Ke^{-\alpha(t-s)} \quad \text{for } t \geq s,$$

$$\|\Phi(t,s)(\mathrm{id}_{\mathbb{R}^d} - P(s))\| \leq Ke^{\alpha(t-s)} \quad \text{for } t \leq s,$$

where $\mathrm{id}_{\mathbb{R}^d}$ [1] denotes the identical linear operator from $\mathbb{R}^d$ into itself. In addition, we say that (1.1) admits an *exponential dichotomy* with growth rate ∞ if there exists a $\gamma \in \mathbb{R}$ such that (1.1) admits an *exponential dichotomy* with growth rate γ and projector $P = \mathrm{id}$, and (1.1) is said to admit an *exponential dichotomy* with growth rate $-\infty$ if there exists a $\gamma \in \mathbb{R}$ such that (1.1) admits an *exponential dichotomy* with growth rate γ and projector $P = 0$, the zero matrix.

Remark 1.1

(i) The range of the projector P of an exponential dichotomy is called a *pseudo-stable space*, and the null space of the projector P is called a *pseudo-unstable space*.
(ii) In the two sided-time case, i.e., $\mathbb{J} = \mathbb{R}$, the pseudo-stable space and pseudo-unstable space are unique. However, in the one sided-time case, $\mathbb{J} = \mathbb{R}_0^+$, in contrast to the pseudo-unstable space, the pseudo-stable space is uniquely determined for exponential dichotomies on $\mathbb{R}_0^+$ [106].

The Sacker–Sell spectrum is then given by set of all growth rates γ such that the linear system does not admit an exponential dichotomy with growth rate γ.

Definition 1.3 (Sacker–Sell Spectrum) The *Sacker–Sell spectrum* (also known as *dichotomy spectrum*) of the linear differential equation (1.1) is defined by

$$\Sigma_{\mathrm{SS}} := \{\gamma \in \overline{\mathbb{R}} : (1.1) \text{ does not admit an exponential dichotomy with growth rate } \gamma\},$$

where $\overline{\mathbb{R}} := \mathbb{R} \cup \{-\infty, \infty\}$.

Example 1.1 If (1.1) is autonomous and γ is not in $\Sigma_{\mathrm{SS}}(A)$, then the projector $P : \mathbb{R} \to \mathbb{R}^{d\times d}$ of an ED of $\dot{x}(t) = [A - \gamma I]x(t)$ is a time-independent projection $P \equiv P(t)$ that yields the decomposition $\mathbb{R}^d = \mathrm{im}\, P \oplus \ker P$ where $\mathrm{im}\, P$ is the sum of all generalized eigenspaces corresponding to those eigenvalues of A with real part less than γ and $\ker P$ is the sum of all generalized eigenspaces corresponding to the eigenvalues of A with real part greater than γ. The Sacker Sell spectrum $\Sigma_{\mathrm{SS}}(A)$ is then given by

[1] Without any confusion, from now on, we use the short notation id instead of $\mathrm{id}_{\mathbb{R}^d}$.

$$\Sigma_{SS}(A) = \{\text{Re}\,\lambda : \lambda \text{ is an eigenvalue of } A\}.$$

The Sacker–Sell spectrum was introduced by R.S. Sacker and G.R. Sell in [108] for skew product flows with compact base. It was generalized to nonautonomous dynamical systems with not necessarily compact base in [12, 112] and for systems defined on a half-line in [106]. For a clearer presentation of the spectral theorem for Sacker–Sell spectrum, in what follows, we consider two separated cases: linear nonautonomous differential equations defined on the half real line and linear nonautonomous differential equations defined on the whole real line.

1.1.2.1 Spectral Theorem of Sacker-Sell Spectrum for Linear Systems on the Half Real Line

The following spectral theorem of Sacker-Sell spectrum for linear systems on the half real line describes the structure of the dichotomy spectrum and the dynamical properties of the solutions starting from the associated filtration, see [83, 106].

Theorem 1.1 (Sacker–Sell Spectral Theorem for Linear Systems on the Half Real Line) *For the linear differential equation* (1.1) *on the half real line* $\mathbb{J} = \mathbb{R}_0^+$, *there exists a* $k \in \{1, \ldots, d\}$ *such that*

$$\Sigma_{SS} = [a_1, b_1] \cup \cdots \cup [a_k, b_k]$$

with $-\infty \le a_1 \le b_1 < a_2 \le b_2 < \cdots < a_k \le b_k \le \infty$. *In addition, there exists a corresponding filtration*

$$\{0\} = \mathcal{W}_0 \subsetneq \mathcal{W}_1 \subsetneq \mathcal{W}_2 \subsetneq \cdots \subsetneq \mathcal{W}_k = \mathbb{R}^d,$$

which satisfies the dynamical characterization

$$\mathcal{W}_i = \left\{\xi \in \mathbb{R}^d : \sup_{t\in\mathbb{R}_0^+} \|X(t)\xi\| e^{-\gamma t} < \infty\right\}$$

for all $i \in \{1, \ldots, k-1\}$ *and* $\gamma \in (b_i, a_{i+1})$.

Note that the linear space $\mathcal{W}_i$ is the pseudo-stable space of the exponential dichotomy with any growth rate taken from the spectral gap interval (b_i, a_{i+1}) for $i \in \{1, \ldots, k-1\}$.

The following result on Sacker–Sell spectra of upper triangular systems follows from [19]. Note that such a statement is only true in the half real line case and does not hold for Sacker–Sell spectra on the whole real line as demonstrated in [19].

Proposition 1.1 (Sacker–Sell Spectrum of Upper Triangular Linear Systems) *Suppose that the linear differential equation* (1.1) *on the half real line* $\mathbb{J} = \mathbb{R}_0^+$ *is upper triangular, i.e.,* $a_{ij}(t) = 0$ *for all* $i > j$ *and* $t \in \mathbb{R}_0^+$, *and assume that* *the off-diagonal elements $a_{ij}(t)$ for all $i < j$ are bounded in $t \in \mathbb{R}_0^+$. Then the Sacker–Sell spectrum of* (1.1) *coincides with that of its diagonal part $\dot{x}_i = a_{ii}(t)x_i$, $i \in \{1, \dots, d\}$, for which the spectrum is the union of the intervals $[\alpha_i, \beta_i]$, where*

$$\begin{aligned} \alpha_i &= \liminf_{t-s\to\infty} \frac{1}{t-s} \int_s^t a_{ii}(u)\, du \\ \textit{and} \quad \beta_i &= \limsup_{t-s\to\infty} \frac{1}{t-s} \int_s^t a_{ii}(u) du \end{aligned} \tag{1.2}$$

for all $i \in \{1, \dots, d\}$.

Remark 1.2 Note that the representation (1.2) does not hold if the diagonal elements of the matrix $A(t)$ are unbounded. As a counterexample, consider the one-dimensional system

$$\dot{x}(t) = a(t)x(t)\,,$$

where $a : \mathbb{R}_0^+ \to \mathbb{R}$ is defined by

$$a(t) = \begin{cases} n & : t \in [2n, 2n+1], n \in \mathbb{N}_0\,, \\ -2n-1 & : t \in [2n+1, 2n+2], n \in \mathbb{N}_0\,. \end{cases}$$

It follows that $\int_n^{n+3} a(u)\, du \le 0$ for all $n \in \mathbb{N}$, and it can be proved that

$$\lim_{t-s\to\infty} \frac{1}{t-s} \int_s^t a(u)\, du = -\infty\,.$$

However, the Sacker–Sell spectrum is given by $[-\infty, \infty]$, since $a(t)$ is arbitrarily close to $-\infty$ and ∞ on intervals of the length one. This shows that the representation (1.2) does not hold for linear systems with unbounded coefficient matrices.

1.1.2.2 Spectral Theorem of Sacker-Sell Spectrum for Linear Systems on the Whole Real Line

When we deal with a linear system on the whole real line, then instead of getting a filtration, we obtain a decomposition associated with the Sacker-Sell spectrum. More concretely, we cite a version of Sacker-Sell spectrum for linear systems on the whole real line in [112, Spectral Theorem] under the assumption that the evolution operator has bounded growth; see [47].

Theorem 1.2 (Sacker–Sell Spectral Theorem for Linear Systems on the Whole Real Line) *Consider the linear differential equation* (1.1) *on the real line $\mathbb{J} = \mathbb{R}$. Suppose that* (1.1) *has bounded growth, i.e., there exist positive constants C, L such* *that*

$$\|\Phi(t,s)\| \leq Ce^{L|t-s|} \qquad \text{for } t,s \in \mathbb{R}.$$

Then, there exists a $k \in \{1,\dots,d\}$ *such that*

$$\Sigma_{SS} = [a_1, b_1] \cup \cdots \cup [a_k, b_k]$$

with $-\infty < a_1 \leq b_1 < a_2 \leq b_2 < \cdots < a_k \leq b_k < \infty$. *In addition, there exists a decomposition*

$$\mathbb{R}^d = \mathcal{W}_1(s) \oplus \cdots \oplus \mathcal{W}_k(s) \qquad \text{for } s \in \mathbb{R},$$

with the property that for any $\varepsilon > 0$, *there exists* $K(\varepsilon) \geq 1$ *such that the following statements hold:*

(i)

$$\|\Phi(t,s)\| \leq K(\varepsilon)e^{(b_k+\varepsilon)(t-s)} \quad \text{for } t \geq s,$$
$$\|\Phi(t,s)\| \leq K(\varepsilon)e^{(a_1-\varepsilon)(t-s)} \quad \text{for } t \leq s.$$

(ii) for $i = 1,\dots,k-1$, *let* $P_i(s)$ *denote the projection onto* $\mathcal{W}_{i+1}(s)$ *through* $\mathcal{W}_i(s)$. *Then,*

$$\|\Phi(t,s)P_i(s)\| \leq K(\varepsilon)e^{(a_{i+1}+\varepsilon)(t-s)} \quad \text{for } t \geq s,$$
$$\|\Phi(t,s)[id - P_i(s)]\| \leq K(\varepsilon)e^{(b_i-\varepsilon)(t-s)} \quad \text{for } t \leq s.$$

1.2 Bohl Spectrum for Nonautonomous Differential Equations

1.2.1 Bohl Spectrum

We first define the Bohl spectrum for each solution of a linear nonautonomous differential equation

$$\dot{x}(t) = A(t)x\,, \qquad t \in \mathbb{R}_0^+, \tag{1.3}$$

where $A : \mathbb{R}_0^+ \to \mathbb{R}^{d\times d}$ is locally integrable. The Bohl spectrum of (1.3) is then the union over the Bohl spectra of the solutions.

Definition 1.4 (Bohl Spectrum) Consider the linear nonautonomous differential equation (1.3) in $\mathbb{R}^d$. The *Bohl spectrum of a particular solution* $X(\cdot)\xi$ of (1.3) is defined as

$$\Sigma_\xi := \Big\{\lambda \in \overline{\mathbb{R}} : \text{there exist sequences } \{t_n\}_{n\in\mathbb{N}} \text{ and } \{s_n\}_{n\in\mathbb{N}}$$
$$\text{with } t_n - s_n \to \infty \text{ such that } \lim_{n\to\infty} \tfrac{1}{t_n - s_n} \ln \tfrac{\|X(t_n)\xi\|}{\|X(s_n)\xi\|} = \lambda\Big\}.$$

The *Bohl spectrum* of (1.3) is defined as

$$\Sigma_{\mathrm{Bohl}} := \bigcup_{\xi\in\mathbb{R}^d\setminus\{0\}} \Sigma_\xi\,.$$

Remark 1.3

(i) By Definition 1.1, we have $\chi_-(\xi), \chi_+(\xi) \in \Sigma_\xi$ for any $\xi \in \mathbb{R}^d\setminus\{0\}$, and we see that in contrast to looking at the asymptotic behavior at infinity of a solution by using the Lyapunov exponent, the Bohl spectrum of this solution provides all possible growth rates of this solution when the length of observation time tends to infinity and the initial time is arbitrary.

(ii) The *upper* and *lower Bohl exponent* of a solution $X(\cdot)\xi$ are defined by

$$\overline{\beta}(\xi) := \limsup_{t-s\to\infty} \frac{1}{t-s} \ln \frac{\|X(t)\xi\|}{\|X(s)\xi\|}, \quad \underline{\beta}(\xi) := \liminf_{t-s\to\infty} \frac{1}{t-s} \ln \frac{\|X(t)\xi\|}{\|X(s)\xi\|},$$

see [54, p. 171–172] and [17]. Thus, $\overline{\beta}(\xi)$ and $\underline{\beta}(\xi)$ measure the biggest and smallest growth rate of the solution $X(\cdot)\xi$, when the length of observation time tends to infinity, and we have

$$\overline{\beta}(\xi) = \sup \Sigma_\xi \quad \text{and} \quad \underline{\beta}(\xi) = \inf \Sigma_\xi\,.$$

(iii) The definition of Bohl spectrum is independent of the norm in $\mathbb{R}^d$.

(iv) A different definition of a Bohl spectrum for discrete systems depending on certain invariant splittings was proposed in [102, Definition 3.8.1], and another spectrum between the Lyapunov and Sacker–Sell spectrum based on nonuniform exponential dichotomies was introduced in [39].

Note that $\overline{\beta}(\xi)$ can be ∞ and $\underline{\beta}(\xi)$ can be $-\infty$. For an arbitrarily chosen $a \in \mathbb{R}$, define

$$[-\infty, a] := (-\infty, a] \cup \{-\infty\}, \qquad [a, \infty] := [a, \infty) \cup \{\infty\}$$

and

$$[-\infty, -\infty] := \{-\infty\}, \qquad [\infty, \infty] := \{\infty\}, \qquad [-\infty, \infty] := \overline{\mathbb{R}}.$$

The following proposition describes fundamental properties of the Bohl spectrum of a particular nonzero solution.

Proposition 1.2 *Consider the linear nonautonomous differential equation* (1.3) *in* $\mathbb{R}^d$. *For all* $\xi \in \mathbb{R}^d \setminus \{0\}$, *the following statements hold:*

(i) We have the representation

$$\Sigma_\xi := \Big\{ \lambda \in \overline{\mathbb{R}} : \text{ there exist sequences } \{t_n\}_{n\in\mathbb{N}} \text{ and } \{s_n\}_{n\in\mathbb{N}} \text{ with } t_n - s_n \to \infty \text{ and } s_n \to \infty \text{ such that } \lim_{n\to\infty} \tfrac{1}{t_n - s_n} \ln \tfrac{\|X(t_n)\xi\|}{\|X(s_n)\xi\|} = \lambda \Big\},$$

i.e., in the definition of the Bohl spectrum, we can always assume $s_n \to \infty$.

(ii) $\Sigma_\xi = \Sigma_{\lambda\xi}$ *for all* $\lambda \in \mathbb{R} \setminus \{0\}$.

(iii) $\Sigma_\xi = \left[\underline{\beta}(\xi), \overline{\beta}(\xi)\right]$.

(iv) Suppose that there exists a constant $M > 0$ *such that*

$$\|A(t)\| \le M \qquad \text{for almost all } t \in \mathbb{R}_0^+ .$$

Then $\Sigma_\xi \subset [-M, M]$.

Proof

(i) Let $\lambda \in \Sigma_\xi$ be arbitrary. Then there exist two sequences $\{t_n\}_{n\in\mathbb{N}}$ and $\{s_n\}_{n\in\mathbb{N}}$ such that $t_n \ge s_n \ge 0$ and

$$\lim_{n\to\infty} t_n - s_n = \infty \quad \text{and} \quad \lim_{n\to\infty} \frac{1}{t_n - s_n} \ln \frac{\|X(t_n)\xi\|}{\|X(s_n)\xi\|} = \lambda .$$

To conclude the proof of this part, we need to construct two sequences $\{\widetilde{t}_n\}_{n\in\mathbb{N}}$ and $\{\widetilde{s}_n\}_{n\in\mathbb{N}}$ such that

$$\lim_{n\to\infty} \widetilde{s}_n = \infty , \quad \lim_{n\to\infty} \widetilde{t}_n - \widetilde{s}_n = \infty ,$$
$$\lim_{n\to\infty} \frac{1}{\widetilde{t}_n - \widetilde{s}_n} \ln \frac{\|X(\widetilde{t}_n)\xi\|}{\|X(\widetilde{s}_n)\xi\|} = \lambda . \tag{1.4}$$

We now consider two separated cases:

Case 1: The sequence $\{s_n\}_{n\in\mathbb{N}}$ is unbounded. Then there exists a subsequence $\{s_{k_n}\}_{n\in\mathbb{N}}$ of $\{s_n\}_{n\in\mathbb{N}}$ such that $\lim_{n\to\infty} s_{k_n} = \infty$. Letting $\widetilde{s}_n := s_{k_n}$ and $\widetilde{t}_n := t_{k_n}$. Then these sequences satisfy (1.4).

Case 2: The sequence $\{s_n\}_{n\in\mathbb{N}}$ is bounded. Let $\Gamma := \sup_{n\in\mathbb{N}} s_n$, and let $n \in \mathbb{N}$ be an arbitrary positive integer. Since $\lim_{m\to\infty} t_m - s_m = \infty$ and

$$\sup_{m\in\mathbb{N}}\left|\ln\frac{\|X(s_m+n)\xi\|}{\|X(s_m)\xi\|}\right| \le \sup_{t\in[0,\Gamma]}\left|\ln\frac{\|X(t+n)\xi\|}{\|X(t)\xi\|}\right| < \infty,$$

it follows that

$$\lim_{m\to\infty}\frac{1}{t_m-s_m-n}\left|\ln\frac{\|X(s_m+n)\xi\|}{\|X(s_m)\xi\|}\right| = 0.$$

Consequently, there exists $k_n \in \mathbb{N}$ such that

$$t_{k_n}-s_{k_n}\ge n^2 \quad\text{and}\quad \frac{1}{t_{k_n}-s_{k_n}-n}\left|\ln\frac{\|X(s_{k_n}+n)\xi\|}{\|X(s_{k_n})\xi\|}\right| \le \frac{1}{n}. \tag{1.5}$$

Define two sequences $\{\widetilde{t}_n\}_{n\in\mathbb{N}}$ and $\{\widetilde{s}_n\}_{n\in\mathbb{N}}$ by

$$\widetilde{t}_n = t_{k_n} \quad\text{and}\quad \widetilde{s}_n := s_{k_n}+n \quad \text{for all } n\in\mathbb{N},$$

where k_n satisfies (1.5). Obviously, $\lim_{n\to\infty}\widetilde{s}_n=\infty$, $\lim_{n\to\infty}\widetilde{t}_n-\widetilde{s}_n=\infty$. It remains to compute $\lim_{n\to\infty}\frac{1}{\widetilde{t}_n-\widetilde{s}_n}\ln\frac{\|X(\widetilde{t}_n)\xi\|}{\|X(\widetilde{s}_n)\xi\|}$. By definition of $\{\widetilde{t}_n\}_{n\in\mathbb{N}}$ and $\{\widetilde{s}_n\}_{n\in\mathbb{N}}$, we have

$$\begin{aligned}\frac{1}{\widetilde{t}_n-\widetilde{s}_n}\ln\frac{\|X(\widetilde{t}_n)\xi\|}{\|X(\widetilde{s}_n)\xi\|} &= \frac{1}{t_{k_n}-s_{k_n}-n}\ln\frac{\|X(t_{k_n})\xi\|}{\|X(s_{k_n}+n)\xi\|}\\ &= \frac{1}{t_{k_n}-s_{k_n}-n}\left(\ln\frac{\|X(t_{k_n})\xi\|}{\|X(s_{k_n})\xi\|}+\ln\frac{\|X(s_{k_n})\xi\|}{\|X(s_{k_n}+n)\xi\|}\right).\end{aligned}$$

Using (1.5), we obtain that

$$\left|\frac{1}{\widetilde{t}_n-\widetilde{s}_n}\ln\frac{\|X(\widetilde{t}_n)\xi\|}{\|X(\widetilde{s}_n)\xi\|}-\frac{1}{t_{k_n}-s_{k_n}-n}\ln\frac{\|X(t_{k_n})\xi\|}{\|X(s_{k_n})\xi\|}\right|\le\frac{1}{n}. \tag{1.6}$$

On the other hand, from $t_{k_n}-s_{k_n}\ge n^2$, we derive that $\lim_{n\to\infty}\frac{t_{k_n}-s_{k_n}}{t_{k_n}-s_{k_n}-n}=1$ and therefore

$$\lim_{n\to\infty}\frac{1}{t_{k_n}-s_{k_n}-n}\ln\frac{\|X(t_{k_n})\xi\|}{\|X(s_{k_n})\xi\|}=\lim_{n\to\infty}\frac{1}{t_{k_n}-s_{k_n}}\ln\frac{\|X(t_{k_n})\xi\|}{\|X(s_{k_n})\xi\|}=\lambda,$$

which together with (1.6) implies that the sequences $\{\widetilde{t}_n\}_{n\in\mathbb{N}}$ and $\{\widetilde{s}_n\}_{n\in\mathbb{N}}$ satisfy (1.4) and the proof of this part is complete.

(ii) This assertion follows directly from Definition 1.4.

(iii) Let $a<b$ be in Σ_ξ, and choose $\lambda\in(a,b)$ arbitrarily. Then there exist sequences $\{t_n\}_{n\in\mathbb{N}}$, $\{s_n\}_{n\in\mathbb{N}}$, $\{\tau_n\}_{n\in\mathbb{N}}$, and $\{\sigma_n\}_{n\in\mathbb{N}}$ such that $t_n-s_n>n$, $\tau_n-\sigma_n>n$,

$$\lim_{n\to\infty} \frac{1}{t_n - s_n} \ln \frac{\|X(t_n)\xi\|}{\|X(s_n)\xi\|} = a \quad \text{and} \quad \lim_{n\to\infty} \frac{1}{\tau_n - \sigma_n} \ln \frac{\|X(\tau_n)\xi\|}{\|X(\sigma_n)\xi\|} = b\,.$$

Consequently, there exists $N \in \mathbb{N}$ such that for all $n \geq N$,

$$\frac{1}{t_n - s_n} \ln \frac{\|X(t_n)\xi\|}{\|X(s_n)\xi\|} < \lambda < \frac{1}{\tau_n - \sigma_n} \ln \frac{\|X(\tau_n)\xi\|}{\|X(\sigma_n)\xi\|}\,. \tag{1.7}$$

Consider the following continuous function $g : [0, 1] \to \mathbb{R}$ defined by

$$g(\theta) := \frac{1}{\theta(t_n - s_n) + (1-\theta)(\tau_n - \sigma_n)} \ln \frac{\|X(\theta t_n + (1-\theta)\tau_n)\xi\|}{\|X(\theta s_n + (1-\theta)\sigma_n)\xi\|}\,.$$

From (1.7), we have $g(0) > \lambda > g(1)$, and by the intermediate value theorem, there exists $\theta_n \in (0, 1)$ such that $g(\theta_n) = \lambda$. This together with the fact that $\lim_{n\to\infty} \theta_n(t_n - s_n) + (1-\theta_n)(\tau_n - \sigma_n) = \infty$ implies that $\lambda \in \Sigma_\xi$ and completes the proof.

(iv) Let $\xi \in \mathbb{R}^d \setminus \{0\}$ be arbitrary. We have the integral equality

$$X(t)\xi = X(s)\xi + \int_s^t A(u)X(u)\xi \, du \quad \text{for all } t \geq s\,.$$

Thus,

$$\|X(t)\xi\| \leq \|X(s)\xi\| + M \int_s^t \|X(u)\xi\| \, du\,.$$

Applying Gronwall's inequality yields that

$$\|X(t)\xi\| \leq e^{M(t-s)} \|X(s)\xi\| \quad \text{for all } t \geq s \geq 0\,,$$

which implies that $\sup \Sigma_\xi \leq M$. Similarly, $\inf \Sigma_\xi \geq -M$. Hence, by Proposition 1.2 (iii), $\Sigma_\xi = [\inf \Sigma_\xi, \sup \Sigma_\xi] \subset [-M, M]$, which completes the proof.

In the following proposition, we show that all linear combination of two solutions satisfying that the angle between them are separated from 0 have the same Bohl spectrum. Recall that the angle between two nonzero vectors $u, v \in \mathbb{R}^d$ is defined by

$$\angle(u, v) := \arccos \frac{|\langle u, v\rangle|}{\|u\|_2 \|v\|_2} \in \left[0, \frac{\pi}{2}\right].$$

Proposition 1.3 *Consider the linear nonautonomous differential equation* (1.1). *Let $x(t)$, $y(t)$ be solutions such that the angle between them is bounded below by a positive number, i.e.,*

$$\inf_{t \in \mathbb{R}_0^+} \angle(x(t), y(t)) > 0.$$

Then if $\alpha\beta \neq 0$, the solutions $t \mapsto \alpha x(t) + \beta y(t)$ all have the same Bohl spectrum.

Proof We use the Euclidean norm $\|\cdot\|$ on $\mathbb{R}^d$. Without loss of generality, we may assume that $\alpha = 1$. So we consider the solutions

$$z(t) = x(t) + \beta y(t).$$

If we define

$$e_1(t) = \frac{x(t)}{\|x(t)\|} \quad \text{and} \quad e_2(t) = \frac{y(t)}{\|y(t)\|},$$

we see that

$$z(t) = \|x(t)\| e_1(t) + \beta \|y(t)\| e_2(t)$$

and

$$\|x(t)\| = (1 - \langle e_1(t), e_2(t)\rangle^2)^{-1}[\langle z(t), e_1(t)\rangle - \langle e_1(t), e_2(t)\rangle \cdot \langle z(t), e_2(t)\rangle]$$

and

$$\beta \|y(t)\| = (1 - \langle e_1(t), e_2(t)\rangle^2)^{-1}[-\langle e_1(t), e_2(t)\rangle \cdot \langle z(t), e_1(t)\rangle + \langle z(t), e_2(t)\rangle].$$

By the angle assumption, we have $1 - \langle e_1(t), e_2(t)\rangle^2 \geq \delta$ for some $\delta > 0$. This implies

$$\|z(t)\| \leq 2 \max\{\|x(t)\|, |\beta| \|y(t)\|\} \leq \frac{4}{\delta} \|z(t)\|.$$

Now let $z_1(t)$ correspond to β_1 and $z_2(t)$ to β_2. Then we note that

$$\frac{\|z_1(t)\|}{\|z_2(t)\|} \leq \frac{4}{\delta} \frac{\max\{\|x(t)\|, |\beta_1| \|y(t)\|\}}{\max\{\|x(t)\|, |\beta_2| \|y(t)\|\}} \leq \frac{4R}{\delta r},$$

where $R = \max\{|\beta_1|, |\beta_2|\}$ and $r = \min\{|\beta_1|, |\beta_2|\}$. Of course, we can interchange the indices 1 and 2 here. Then

$$\frac{\|z_1(t)\|}{\|z_1(s)\|} = \frac{\|z_1(t)\|}{\|z_2(t)\|} \frac{\|z_2(t)\|}{\|z_2(s)\|} \frac{\|z_2(s)\|}{\|z_1(s)\|} \leq \frac{16R^2}{\delta^2 r^2} \frac{\|z_2(t)\|}{\|z_2(s)\|} = N \frac{\|z_2(t)\|}{\|z_2(s)\|}.$$

It follows that

$$\frac{1}{t-s}\ln\frac{\|z_1(t)\|}{\|z_1(s)\|} \le \frac{\ln N}{t-s} + \frac{1}{t-s}\ln\frac{\|z_2(t)\|}{\|z_2(s)\|}.$$

Thus,

$$\limsup_{t-s\to\infty}\frac{1}{t-s}\ln\frac{\|z_1(t)\|}{\|z_1(s)\|} \le \limsup_{t-s\to\infty}\frac{1}{t-s}\ln\frac{\|z_2(t)\|}{\|z_2(s)\|}.$$

Switching the indices 1 and 2, we get equality. Next from

$$\frac{1}{t-s}\ln\frac{\|z_2(t)\|}{\|z_2(s)\|} \ge -\frac{\ln N}{t-s} + \frac{1}{t-s}\ln\frac{\|z_1(t)\|}{\|z_1(s)\|},$$

we get

$$\liminf_{t-s\to\infty}\frac{1}{t-s}\ln\frac{\|z_2(t)\|}{\|z_2(s)\|} \ge \liminf_{t-s\to\infty}\frac{1}{t-s}\ln\frac{\|z_1(t)\|}{\|z_1(s)\|}$$

and switching the indices 1 and 2, we get equality also. The conclusion is that $z_1(t)$ and $z_2(t)$ have the same Bohl spectrum.

Remark 1.4 We demonstrate that the common Bohl spectrum of the solution $t \mapsto \alpha x(t) + \beta y(t)$ in Proposition 1.3 does not depend just on Σ_x and Σ_y. Consider the diagonal system

$$\dot{x}(t) = 0,$$
$$\dot{y}(t) = a(t)y(t),$$

where $a(t) = 1$ if $T_{2k} \le t \le T_{2k+1}$, and $a(t) = -1$ if $T_{2k+1} \le t \le T_{2k+2}$. Here T_k is an increasing sequence with $T_0 = 0$ and $T_{k+1} - T_k \to \infty$ as $k \to \infty$. Then if we take the solutions $x(t) = (1, 0)$ and $y(t) = \big(0, \exp(\int_0^t a(u)\,du)\big)$, it is easy to see that $\Sigma_x = \{0\}$ and $\Sigma_y = [-1, 1]$. By appropriate choice of the sequence T_k, we can arrange that $\int_0^t a(u)\,du \ge 0$ for $t \ge 0$. Then if we use the maximum norm in $\mathbb{R}^2$, we see that $\|x(t) + y(t)\| = |y(t)|$ for all $t \ge 0$. This means that for all $t \ge s \ge 0$, we have

$$\frac{1}{t-s}\ln\frac{\|x(t) + y(t)\|}{\|x(s) + y(s)\|} = \frac{1}{t-s}\ln\frac{\|y(t)\|}{\|y(s)\|}.$$

It follows that $\Sigma_{x+y} = \Sigma_y$.

On the other hand, again by appropriate choice of the sequence T_k, we can arrange that $\int_0^t a(u)\,du \le 0$ for $t \ge T_2$. Then if we use the maximum norm in $\mathbb{R}^2$, we see that $\|x(t) + y(t)\| = |x(t)| = 1$ for all $t \ge T_2$. So for all $t \ge T_2$ and $s \ge T_2$, we get

$$\frac{1}{t-s}\ln\frac{\|x(t)+y(t)\|}{\|x(s)+y(s)\|}=\frac{1}{t-s}\ln\frac{1}{1}=0,$$

which implies that $\Sigma_{x+y}=\Sigma_x$.

1.2.2 Bohl Spectral Theorem

We prove in this section that the Bohl spectrum of a locally integrable linear nonautonomous differential equation consists of at most finitely many intervals, the number of which is bounded by the dimension of the system, and we associate a filtration of subspaces to these spectral intervals.

Theorem 1.3 (Bohl Spectral Theorem) *Consider the linear nonautonomous differential equation* (1.1) *in* $\mathbb{R}^d$. *Then its Bohl spectrum consists of k (not necessarily closed) disjoint intervals, i.e.,*

$$\Sigma_{\mathrm{Bohl}}=I_1\cup\cdots\cup I_k,$$

where $1\le k\le d$ *and* $I_1, I_2,\dots,I_k$ *are ordered intervals. The interval* I_1 *can be unbounded from below,* I_k *can be unbounded from above, and* $I_2,\dots,I_{k-1}$ *are bounded. There exists a corresponding filtration*

$$\{0\}=\mathcal{S}_0\subsetneq\mathcal{S}_1\subsetneq\mathcal{S}_2\subsetneq\cdots\subsetneq\mathcal{S}_k=\mathbb{R}^d \tag{1.8}$$

satisfying the following dynamical characterization

$$\mathcal{S}_i\setminus\{0\}=\left\{\xi\in\mathbb{R}^d:\Sigma_\xi\subset\bigcup_{j=1}^{i}I_j\right\}\quad \textit{for all}\ i\in\{1,\dots,k\}. \tag{1.9}$$

Proof Let $\lambda\in\overline{\mathbb{R}}\setminus\Sigma_{\mathrm{Bohl}}$ be arbitrary. Due to Proposition 1.2 (iii), for any $\xi\in\mathbb{R}^d\setminus\{0\}$, either $\Sigma_\xi\subset[-\infty,\lambda)$ or $\Sigma_\xi\subset(\lambda,+\infty]$ holds. Define

$$\mathcal{M}_\lambda:=\left\{\xi\in\mathbb{R}^d\setminus\{0\}:\Sigma_\xi\subset[-\infty,\lambda)\right\}\cup\{0\} \tag{1.10}$$

and

$$\mathcal{N}_\lambda:=\left\{\xi\in\mathbb{R}^d\setminus\{0\}:\Sigma_\xi\subset(\lambda,\infty]\right\}.$$

Obviously, $\mathcal{M}_\lambda\cup\mathcal{N}_\lambda=\mathbb{R}^d$, and we show now that $\mathcal{M}_\lambda$ is a linear subspace of $\mathbb{R}^d$. Consider $\xi,\eta\in\mathcal{M}_\lambda$ and $\alpha,\beta\in\mathbb{R}$ with $\alpha\xi+\beta\eta\neq 0$. Since $\Sigma_\xi,\Sigma_\eta\subset[-\infty,\lambda)$, it follows that

$$\limsup_{t\to\infty}\frac{1}{t}\ln\|X(t)\xi\|<\lambda\quad\text{and}\quad\limsup_{t\to\infty}\frac{1}{t}\ln\|X(t)\eta\|<\lambda,$$

which implies that there exist $K > 0$ and $\mu < \lambda$ such that

$$\max\{\|X(t)\xi\|, \|X(t)\eta\|\} \le K e^{\mu t} \quad \text{for all } t \ge 0 .$$

Consequently,

$$\|X(t)(\alpha\xi + \beta\eta)\| \le K(|\alpha| + |\beta|)e^{\mu t} \quad \text{for all } t \ge 0 .$$

Hence, there exists a sequence $\{t_n\}_{n\in\mathbb{N}}$ tending to infinity with

$$\lim_{n\to\infty} \frac{1}{t_n} \ln \|X(t_n)(\alpha\xi + \beta\eta)\| \in [-\infty, \lambda) .$$

Thus, $\Sigma_{\alpha\xi+\beta\eta} \cap [-\infty, \lambda) \neq \emptyset$, and since $\Sigma_{\alpha\xi+\beta\eta}$ is an interval that does not contain λ, it must be a subset of $[-\infty, \lambda)$, and thus, we have $\alpha\xi + \beta\eta \in \mathcal{M}_\lambda$. Hence, $\mathcal{M}_\lambda$ is a linear subspace of $\mathbb{R}^d$.

Let $d_0 < d_1 < \cdots < d_n$ be elements of the set $\{\dim(\mathcal{M}_\lambda) : \lambda \in \overline{\mathbb{R}} \setminus \Sigma_{\mathrm{Bohl}}\}$. Depending on whether $\pm\infty \in \Sigma_{\mathrm{Bohl}}$ or not, we have the following estimate on the number n:

(a) If $\pm\infty \in \Sigma_{\mathrm{Bohl}}$, then $d_0 \ge 1$ and $d_n \le d - 1$ and therefore $n \le d - 2$.
(b) If $-\infty \in \Sigma_{\mathrm{Bohl}}$ and $+\infty \notin \Sigma_{\mathrm{Bohl}}$, then $d_0 \ge 1$ and therefore $n \le d - 1$.
(c) If $-\infty \notin \Sigma_{\mathrm{Bohl}}$ and $+\infty \in \Sigma_{\mathrm{Bohl}}$, then $d_n \le d - 1$ and therefore $n \le d - 1$.
(d) If $-\infty \notin \Sigma_{\mathrm{Bohl}}$ and $+\infty \notin \Sigma_{\mathrm{Bohl}}$, then $n \le d$.

For $i \in \{0, \ldots, n\}$, we define

$$J_i := \left\{\lambda \in \overline{\mathbb{R}} \setminus \Sigma_{\mathrm{Bohl}} : \dim(\mathcal{M}_\lambda) = d_i\right\}.$$

We now show that each set J_i is an interval. Let $i \in \{0, \ldots, n\}$ and $a < b$ be two elements of J_i. We show now that $[a, b] \subset J_i$. By (1.10), we have $\mathcal{M}_a \subset \mathcal{M}_b$, and using $\dim(\mathcal{M}_a) = \dim(\mathcal{M}_b)$, this implies that $\mathcal{M}_a = \mathcal{M}_b$, and thus $\mathcal{N}_a = \mathcal{N}_b$. This means that $\mathbb{R}^d = \mathcal{M}_a \cup \mathcal{N}_b$. Let $\lambda \in [a, b]$ be arbitrary and $\xi \in \mathbb{R}^d \setminus \{0\}$. Thus, either $\xi \in \mathcal{M}_a$ or $\xi \in \mathcal{N}_b$. In both of these cases, we have $\lambda \notin \Sigma_\xi$ and therefore $\lambda \in \overline{\mathbb{R}} \setminus \Sigma_{\mathrm{Bohl}}$. Now, we know that $\mathcal{M}_\lambda$ is a linear subspace, and by (1.10), we have $\mathcal{M}_a \subset \mathcal{M}_\lambda \subset \mathcal{M}_b$. Thus, $\mathcal{M}_a = \mathcal{M}_\lambda = \mathcal{M}_b$ and therefore $\lambda \in J_i$. This means that we have proved that J_i is an interval. Obviously, the order of the intervals $J_0, \ldots, J_n$ is $J_0 < J_1 < \cdots < J_n$, and we have

$$\Sigma_{\mathrm{Bohl}} = \overline{\mathbb{R}} \setminus \bigcup_{i=0}^{n} J_i .$$

Let k denote the number of disjoint intervals I_i of Σ_{Bohl}. According to the cases (a)–(d) above, we have the following dependence of k and n:

(i) $k = n + 2$ in case (a) above,

(ii) $k = n + 1$ in case (b) and (c) above,

(iii) $k = n$ in case (d) above.

Thus, from the relation between n and d established above, we always obtain that $k \leq d$. To conclude the proof, for each $i \in \{1, \dots, k\}$, we define the set $\mathcal{S}_i$ as in (1.9) together with $\{0\}$. Note that the space $\mathcal{S}_i$ coincides with $\mathcal{M}_\lambda$ for $\lambda = \frac{1}{2}(\sup I_i + \inf I_{i+1})$, where $i \in \{1, \dots, k-1\}$, and $\mathcal{S}_k = \mathcal{M}_\lambda = \mathbb{R}^d$ for $\lambda > \sup I_k$. Then, clearly $\mathcal{S}_i$ is a linear subspace and satisfies (1.8). This finishes the proof.

Next, we concentrate on constructing an example of a nonautonomous differential equation such that its Bohl spectrum is not closed. Our construction is implicit by using a result from [16]:

Let $\mathcal{M}_d$ denote the set of all piecewise continuous and uniformly bounded matrix-valued functions $A : \mathbb{R}_0^+ \to \mathbb{R}^{d\times d}$. For each $A \in \mathcal{M}_d$, consider the linear nonautonomous differential equation

$$\dot{x}(t) = A(t)x(t)\,. \tag{1.11}$$

Consider the *uniform upper exponent function* of (1.11), $\overline{\beta}_A : \mathbb{R}^d \setminus \{0\} \to \mathbb{R}$, where $\overline{\beta}_A(\xi)$ is the upper Bohl exponent of the solution $X(t)\xi$ of (1.11); see Remark 1.3(ii). A complete description of the set of functions $\overline{\mathcal{B}}_d := \{\overline{\beta}_A : A \in \mathcal{M}_d\}$ is given as follows (see [16, Theorem 1]).

Theorem 1.4 *A function $\beta : \mathbb{R}^d \setminus \{0\} \to \mathbb{R}$ belongs to the class $\overline{\mathcal{B}}_d$ if and only if it satisfies the following three conditions:*

(i) β is bounded,

(ii) $\beta(\xi) = \beta(\alpha\xi)$ for any nonzero $\alpha \in \mathbb{R}$ and $\xi \in \mathbb{R}^d \setminus \{0\}$,

(iii) For any $q \in \mathbb{R}$, the set $\{\xi : \beta(\xi) \geq q\}$ is a G_δ set.

The following example shows that the intervals of the Bohl spectrum do not need to be closed.

Example 1.2 Consider the function $\beta : \mathbb{R}^2 \setminus \{0\} \to \mathbb{R}$ defined by

$$\beta(r\cos\varphi, r\sin\varphi) = \begin{cases} 0 & : \varphi = 0, \\ |\cos\varphi| & : \varphi \in (0, 2\pi), \end{cases}$$

where $r \in (0, \infty)$. Obviously, the function β satisfies the three conditions of Theorem 1.4. Consequently, there exists a piecewise continuous and uniformly bounded matrix-valued function $A : \mathbb{R}_0^+ \to \mathbb{R}^{2\times 2}$ such that $\overline{\beta}_A \equiv \beta$. By construction of β_A, it is easy to see that $[0, 1) \subset \Sigma_{\mathrm{Bohl}}$. Suppose to the contrary that Σ_{Bohl} is closed. Thus, $1 \in \Sigma_{\mathrm{Bohl}}$, which means that there exists $\xi \in \mathbb{R}^2 \setminus \{0\}$ such that $1 \in \Sigma_\xi$. That leads to a contradiction, since $\overline{\beta}_A(\xi) < 1$. Thus, Σ_{Bohl} is not closed.

In the remaining part of this section, we show that Bohl spectrum is preserved under a kinematic similarity transformation. Recall that a linear nonautonomous differential equation

$$\dot{x}(t) = A(t)x(t) \quad \text{for all } t \in \mathbb{R}_0^+ \tag{1.12}$$

is said to be *kinematically similar* to another linear nonautonomous differential equation

$$\dot{y} = B(t)y \quad \text{for all } t \in \mathbb{R}_0^+ \tag{1.13}$$

if there exists a continuously differentiable function $S : \mathbb{R}_0^+ \to \mathbb{R}^{d\times d}$ of invertible matrices such that both S and S^{-1} are bounded, and which satisfies the differential equation

$$\dot{S}(t) = A(t)S(t) - S(t)B(t) \quad \text{for all } t \in \mathbb{R}_0^+ \tag{1.14}$$

(see [47, p. 38]).

Proposition 1.4 (Invariance of the Bohl Spectrum Under Kinematic Similarity Transformations) *Suppose that* (1.12) *and* (1.13) *are kinematically similar. Then the Bohl spectra* $\Sigma_{\text{Bohl}}(A)$ *of* (1.12) *and* $\Sigma_{\text{Bohl}}(B)$ *of* (1.13) *coincide.*

Proof Let $X_A(t)$ and $X_B(t)$ denote the fundamental matrix solution of (1.12) and (1.13), respectively. From (1.14), we derive

$$X_B(t) = S(t)^{-1}X_A(t)S(0) \quad \text{for all } t \geq 0,$$

which implies that the Bohl spectrum of the solution $X_A(t)S(0)\xi$ of (1.12) is equal to the Bohl spectrum of the solution $X_B(t)\xi$ for all $\xi \in \mathbb{R}^d \setminus \{0\}$, where we use the inequality

$$\ln\|y\| - \ln\|S^{-1}(t)\| \leq \ln\|S(t)y\| \leq \ln\|y\| + \ln\|S(t)\|.$$

Since $S(0)$ is invertible, it follows that $\Sigma_{\text{Bohl}}(A) = \Sigma_{\text{Bohl}}(B)$ and the proof is complete.

1.2.3 A Comparison of Bohl and Sacker–Sell Spectrum

This subsection is devoted to a comparison of the Bohl spectrum with the Sacker–Sell spectrum. Since the Sacker–Sell spectrum consists of closed intervals, it follows directly from Example 1.2 that the Bohl spectrum does not always coincide with the Sacker–Sell spectrum. In this section, we show that the Bohl spectrum is always a subset of the Sacker–Sell spectrum, and we provide sufficient conditions under which both spectra coincide. The sufficient conditions are

(c1) the systems are diagonalizable,

(c2) the systems are bounded integrally separated

(c3) the Sacker-Sell spectra intervals are singleton.

We now recall the notion of integral separation of linear nonautonomous differential equations; see e.g., [1, Definition 5.3.1].

Definition 1.5 (Integral Separation of Linear Nonautonomous Differential Equations) Let $\xi, \eta \in \mathbb{R}^d \setminus \{0\}$. Then the two solutions $X(t)\xi$ and $X(t)\eta$ of (1.1) are said to be *integrally separated* if there exists $K \geq 1$ and $\alpha > 0$ such that

$$\frac{\|X(t)\xi\|}{\|X(s)\xi\|} \geq Ke^{\alpha(t-s)} \frac{\|X(t)\eta\|}{\|X(s)\eta\|} \quad \text{for all } t \geq s \geq 0. \tag{1.15}$$

The linear nonautonomous differential equation (1.1) is said to be *integrally separated* if there are d independent solutions $X(t)\xi_1, \ldots, X(t)\xi_d$ such that $X(t)\xi_i$ and $X(t)\xi_{i+1}$ are integrally separated for all $i \in \{1, \ldots, d-1\}$.

In the next lemma, we show that when the solutions are integrally separated and $X(t)\xi$ is the bigger solution in the above sense, then the Bohl spectrum of any nontrivial linear combination of $X(t)\xi$ and $X(t)\eta$ is always given by Σ_ξ.

Lemma 1.1 *Consider $\xi, \eta \in \mathbb{R}^d \setminus \{0\}$ such that the two solutions $X(t)\xi$ and $X(t)\eta$ of* (1.1) *are integrally separated, i.e., the inequality* (1.15) *holds. Then*

$$\Sigma_{\lambda\xi+\mu\eta} = \Sigma_\xi \quad \textit{for all } \lambda \in \mathbb{R} \setminus \{0\} \textit{ and } \mu \in \mathbb{R}. \tag{1.16}$$

Proof The lemma is clear for $\mu = 0$. For the rest, we may prove (1.16) for the case that $\lambda = \mu = 1$. By taking $s = 0$ in (1.15), there exists $T > 0$ such that

$$\frac{\|X(t)\eta\|}{\|X(t)\xi\|} \leq \frac{1}{2} \quad \text{for all } t \geq T.$$

Thus, for all $t \geq T$, we have

$$\frac{1}{2} \leq 1 - \frac{\|X(t)\eta\|}{\|X(t)\xi\|} \leq \frac{\|X(t)(\xi+\eta)\|}{\|X(t)\xi\|} \leq 1 + \frac{\|X(t)\eta\|}{\|X(t)\xi\|} \leq \frac{3}{2}. \tag{1.17}$$

By (1.17), for all $t \geq s \geq T$, we have

$$\frac{\|X(t)(\xi+\eta)\|}{\|X(s)(\xi+\eta)\|} = \frac{\|X(t)(\xi+\eta)\|}{\|X(t)\xi\|} \frac{\|X(t)\xi\|}{\|X(s)\xi\|} \frac{\|X(s)\xi\|}{\|X(s)(\xi+\eta)\|} \leq 3 \frac{\|X(t)\xi\|}{\|X(s)\xi\|},$$

which implies that

$$\frac{1}{t-s}\ln\frac{\|X(t)(\xi+\eta)\|}{\|X(s)(\xi+\eta)\|} \le \frac{\ln 3}{t-s} + \frac{1}{t-s}\ln\frac{\|X(t)\xi\|}{\|X(s)\xi\|}. \tag{1.18}$$

Conversely, for all $t \ge s \ge T$, we have

$$\frac{\|X(t)(\xi+\eta)\|}{\|X(s)(\xi+\eta)\|} = \frac{\|X(t)(\xi+\eta)\|}{\|X(t)\xi\|}\frac{\|X(t)\xi\|}{\|X(s)\xi\|}\frac{\|X(s)\xi\|}{\|X(s)(\xi+\eta)\|} \ge \frac{1}{3}\frac{\|X(t)\xi\|}{\|X(s)\xi\|},$$

so that

$$\frac{1}{t-s}\ln\frac{\|X(t)(\xi+\eta)\|}{\|X(s)(\xi+\eta)\|} \ge -\frac{\ln 3}{t-s} + \frac{1}{t-s}\ln\frac{\|X(t)\xi\|}{\|X(s)\xi\|}. \tag{1.19}$$

Let $\{t_n\}_{n\in\mathbb{N}}$ and $\{s_n\}_{n\in\mathbb{N}}$ be two positive sequences with $\lim_{n\to\infty}(t_n - s_n) = \infty$ and $\lim_{n\to\infty} s_n = \infty$. Since $\lim_{n\to\infty} t_n = \lim_{n\to\infty} s_n = \infty$, there exists $N \in \mathbb{N}$ such that $t_n \ge s_n \ge T$ for all $n \ge N$. Hence, combining (1.18) and (1.19) yields

$$\lim_{n\to\infty}\frac{1}{t_n - s_n}\ln\frac{\|X(t_n)(\xi+\eta)\|}{\|X(s_n)(\xi+\eta)\|} = \lim_{n\to\infty}\frac{1}{t_n - s_n}\ln\frac{\|X(t_n)\xi\|}{\|X(s_n)\xi\|},$$

whenever one of the two above limits exists. This fact, together with Lemma 1.2 (i), shows that $\Sigma_{\xi+\eta} = \Sigma_\xi$. This concludes the proof of this lemma.

We first use this lemma to show that the Bohl spectrum is a subset of the Sacker–Sell spectrum. As a consequence, the filtration corresponding to the Bohl spectrum is finer than the filtration corresponding to Sacker–Sell spectrum.

Theorem 1.5 *Consider the Bohl spectrum Σ_{Bohl} and the Sacker–Sell spectrum Σ_{SS} of a linear nonautonomous differential equation* (1.1). *The following statements hold:*

(i) The Bohl spectrum is a subset of the Sacker–Sell spectrum.
(ii) The filtration associated with the Bohl spectrum is finer than the one of Sacker–Sell spectrum.

Proof

(i) Let $\lambda \in \mathbb{R} \setminus \Sigma_{\mathrm{SS}}$ be arbitrary. Then $\dot{x}(t) = A(t)x(t)$ has an exponential dichotomy with growth rate λ, which means that there exists a projector $P \in \mathbb{R}^{d\times d}$ such that

$$\|X(t)PX^{-1}(s)\| \le Ke^{(\lambda-\alpha)(t-s)} \quad \text{for all } 0 \le s \le t \tag{1.20}$$

and

$$\|X(t)(\mathrm{id}-P)X^{-1}(s)\| \le Ke^{-(\lambda+\alpha)(s-t)} \quad \text{for all } 0 \le t \le s. \tag{1.21}$$

Then for any $\xi \in \ker P \setminus \{0\}$ and $\eta \in \operatorname{im} P \setminus \{0\}$, the solutions $X(t)\xi$ and $X(t)\eta$ are integrally separated. So, by virtue of Lemma 1.1, we have

$$\Sigma_{\text{Bohl}} = \bigcup_{\xi \in \ker P\setminus\{0\}} \Sigma_\xi \cup \bigcup_{\eta \in \operatorname{im} P\setminus\{0\}} \Sigma_\eta .$$

From (1.20), we derive that for all $\eta \in \operatorname{im} P \setminus \{0\}$,

$$\|X(t)\eta\| = \|X(t)P\eta\| \leq Ke^{(\lambda-\alpha)(t-s)}\|X(s)\eta\| ,$$

which implies that $\Sigma_\eta \subset (-\infty, \lambda - \alpha]$. Similarly, using (1.21), we obtain that if ξ is in the kernel of P, then $\Sigma_\xi \subset [\lambda + \alpha, \infty)$. Thus, $\lambda \notin \Sigma_{\text{Bohl}}$, which finishes the proof of (i).

(ii) Let I_j be the rightmost component of the Bohl spectrum contained in Sacker–Sell spectral interval $[a_i, b_i]$, and let $\mathcal{S}_j$ be the subspace in the Bohl filtration corresponding to the union of I_j and the intervals to its left. Then $\mathcal{S}_j = \mathcal{M}_\lambda = \left\{\xi \in \mathbb{R}^d : \Sigma_\xi \subset [-\infty, \lambda)\right\}$ for $\lambda \in (b_i, a_{i+1})$, where $\mathcal{M}_\lambda$ is defined as in (1.10). If $\xi \in S_j$, then for each $\lambda \in (b_i, a_{i+1})$, there exists $K \geq 1$ such that

$$\|X(t)\xi\| \leq Ke^{\lambda(t-s)}\|X(s)\xi\| \quad \text{for all } t \geq s .$$

It follows that ξ is in the pseudo-stable subspace for the exponential dichotomy with growth rate λ of $\dot{x}(t) = A(t)x(t)$ for $\lambda \in (b_i, a_{i+1})$. Hence, $\xi \in \mathcal{W}_i$, which proves $\mathcal{S}_j \subset \mathcal{W}_i$, with $\mathcal{W}_i$ defined as in Theorem 1.1. Conversely, note that since $\mathcal{W}_i$ is the pseudo-stable subspace for the exponential dichotomy with growth rate λ of $\dot{x}(t) = A(t)x(t)$ for $\lambda \in (b_i, a_{i+1})$, there exist constants $K, \alpha > 0$ such that for all $\xi \in \mathcal{W}_i$, we have

$$\|X(t)\xi\| \leq Ke^{(\lambda-\alpha)(t-s)}\|X(s)\xi\| \quad \text{for all } t \geq s .$$

This means that $\Sigma_\xi \subset (-\infty, \lambda)$ for all $\lambda \in (b_i, a_{i+1})$ which implies that $\xi \in \mathcal{S}_j$. Thus $\mathcal{S}_j = \mathcal{W}_i$. In fact, what we have proved is that the Bohl filtration is $\mathcal{S}_i$, $i \in \{1, \ldots, m\}$, and the Sacker–Sell filtration is $\mathcal{W}_i$, for $i \in \{1, \ldots, n\}$, where $n \leq m \leq d$, and there exist $i_1 < i_2 < \cdots < i_n$ such that $\mathcal{S}_{i_j} = \mathcal{W}_j$.

Now, we prove that for diagonalizable systems, Bohl and Sacker–Sell coincide.

Theorem 1.6 (Bohl and Sacker–Sell Spectra of Diagonalizable Systems) *Suppose that the linear nonautonomous differential equation* (1.1) *is* diagonalizable, *i.e., it is kinematically similar to a (nonautonomous) diagonal system. Then the Bohl and Sacker–Sell spectrum of* (1.1) *coincide. In particular, both spectra coincide for one-dimensional systems.*

Proof By assumption, the linear system (1.1) is kinematically similar to a diagonal system

$$\dot{x}(t) = D(t)x(t) = \mathrm{diag}(a_1(t), \dots, a_d(t))\, x(t)\,. \tag{1.22}$$

Since the Bohl and Sacker–Sell spectra are invariant under kinematic similarity, it is sufficient to show that the Bohl spectrum Σ_{Bohl} and the dichotomy spectrum Σ_{SS} of (1.22) coincide. For $i \in \{1, \dots, d\}$, define

$$\alpha_i := \liminf_{t-s\to\infty} \frac{1}{t-s}\int_s^t a_i(u)\,\mathrm{d}u \quad \text{and} \quad \beta_i = \limsup_{t-s\to\infty} \frac{1}{t-s}\int_s^t a_i(u)\,\mathrm{d}u\,.$$

It follows from Proposition 1.1 that

$$\Sigma_{\mathrm{SS}} = \bigcup_{i=1}^{d} [\alpha_i, \beta_i]\,.$$

To compute Σ_{Bohl}, let $(e_1, \dots, e_d)$ denote the standard orthonormal basis of $\mathbb{R}^d$. A simple computation yields that

$$\Sigma_{e_i} = [\alpha_i, \beta_i] \quad \text{for all } i \in \{1, \dots, d\}\,,$$

which implies

$$\bigcup_{i=1}^{d}[\alpha_i, \beta_i] \subset \Sigma_{\mathrm{Bohl}} \subset \Sigma_{\mathrm{SS}} = \bigcup_{i=1}^{d}[\alpha_i, \beta_i]\,,$$

and completes the proof.

We now prove using the previous theorem that the Bohl and Sacker–Sell spectra coincide for bounded integrally separated systems. This means also that the Bohl spectrum depends continuously on parameters for such systems.

Corollary 1.1 *Suppose that system* (1.1) *is integrally separated and* $A(t)$ *is bounded in* $t \in \mathbb{R}_0^+$. *Then the Bohl spectrum coincides with the Sacker–Sell spectrum of* (1.1).

Proof By Bylov's Theorem [1, Theorem 5.3.1, p. 149], the linear system (1.1) is kinematically similar to a diagonal system

$$\dot{x}(t) = D(t)x(t) = \mathrm{diag}(a_1(t), \dots, a_d(t))x(t)\,,$$

and a direct application of Theorem 1.6 completes the proof.

Remark 1.5 The boundedness assumption of $A(t)$ in the above corollary is needed, since there exists an unbounded integrally separated system which is not diagonalizable such that its Bohl spectrum and its Sacker-Sell spectrum are different. Consider the system $\dot{x}(t) = A(t)x(t)$, where $A(t)$ is defined by

$$A(t) := \begin{pmatrix} 0 & 2e^t \\ 0 & 1 \end{pmatrix} \quad \text{for all } t \geq 0 .$$

The fundamental matrix solution $X(t)$ of this system is given by

$$X(t) = \begin{pmatrix} 1 & e^{2t} - 1 \\ 0 & e^t \end{pmatrix} \quad \text{for all } t \geq 0 .$$

Note that

$$X(t)\begin{pmatrix} 1 \\ 0 \end{pmatrix} = \begin{pmatrix} 1 \\ 0 \end{pmatrix} \quad \text{and} \quad X(t)\begin{pmatrix} 0 \\ 1 \end{pmatrix} = \begin{pmatrix} e^{2t} - 1 \\ e^t \end{pmatrix},$$

which implies that these two solutions are integrally separated. By explicit presentation of $X(t)$ given above, we can easily see that $\Sigma_{\mathrm{SS}} = [0, 2]$ and $\Sigma_{\mathrm{Bohl}} = \{0\} \cup \{2\}$.

Let $\mathcal{B}$ denote the linear space of bounded measurable matrix-valued functions $A : \mathbb{R}_0^+ \to \mathbb{R}^{d \times d}$. We endow $\mathcal{B}$ with the L^∞-norm defined by

$$\|A - B\|_\infty = \operatorname{ess\,sup}_{t \in \mathbb{R}_0^+} \|A(t) - B(t)\| ,$$

so that $(\mathcal{B}, \|\cdot\|_\infty)$ is a Banach space. Using [92], one can show that there exists an open and dense set $\mathcal{R}$ of $\mathcal{B}$ such that for all $A \in \mathcal{R}$, the associated linear nonautonomous differential equation is integrally separated (note that genericity of exponential dichotomies for two-dimensional quasiperiodic linear systems was treated in [62]). As a consequence, we obtain the following corollary.

Corollary 1.2 (Coincidence Is Generic) *The Bohl spectrum and the Sacker–Sell spectrum coincide generically for bounded linear nonautonomous differential equations.*

Suppose the Sacker–Sell spectrum consists of points. Then by Theorem 1.5, the Bohl spectrum consists of points. We still need to prove each point in the Sacker–Sell spectrum is also in the Bohl spectrum. This follows from the next lemma.

Lemma 1.2 *Let* $[a, b]$ *be a spectral interval of the Sacker–Sell spectrum of the linear nonautonomous differential equation* (1.1). *Then there exists a solution whose Bohl spectrum is contained in* $[a, b]$.

Proof Let $\{I_i = [a_i, b_i]\}_{i \in \{1,\dots,k\}}$ be the ordered Sacker–Sell spectral intervals of (1.1). Consider the filtration

$$\{0\} = \mathcal{W}_0 \subsetneq \mathcal{W}_1 \subsetneq \mathcal{W}_2 \subsetneq \cdots \subsetneq \mathcal{W}_k = \mathbb{R}^d ,$$

established in Theorem 1.1, satisfying the dynamical characterization

$$\mathcal{W}_i = \left\{\xi \in \mathbb{R}^d : \sup_{t\in\mathbb{R}_0^+} \|X(t)\xi\|e^{-\gamma t} < \infty\right\}$$

for all $i \in \{1, \dots, k-1\}$ and $\gamma \in (b_i, a_{i+1})$. There exists an $i \in \{1, \dots, k\}$ such that $I_i = [a, b] = [a_i, b_i]$. Note that $\mathcal{W}_{i-1}$ is a proper subspace of $\mathcal{W}_i$ and we can write $\mathcal{W}_i = \mathcal{W}_{i-1} \oplus \mathcal{V}$ with $\mathcal{V} \neq \{0\}$. Now $\dot{x}(t) = A(t)x(t)$ has an exponential dichotomy with growth rate $b+\varepsilon$ with pseudo-stable subspace $\mathcal{W}_i$. This means that for all $\xi \in \mathcal{W}_i$, there exist $K_1 > 0$ and $\alpha_1 > 0$ such that

$$\frac{\|X(t)\xi\|}{\|X(s)\xi\|} \leq K_1 e^{(b+\varepsilon-\alpha_1)(t-s)} \quad \text{for all } t \geq s \geq 0\,. \tag{1.23}$$

Next $\dot{x}(t) = A(t)x(t)$ has an exponential dichotomy with growth rate $a - \varepsilon$ with a pseudo-unstable subspace $\mathcal{V}$ [106, Remark 5.6 and Lemma 6.1]. This means that for all $\xi \in \mathcal{V}$, there exist $K_2 > 0$ and $\alpha_2 > 0$ such that

$$\frac{\|X(t)\xi\|}{\|X(s)\xi\|} \geq K_2 e^{(a-\varepsilon+\alpha_2)(t-s)} \quad \text{for all } t \geq s \geq 0\,. \tag{1.24}$$

From (1.23), it follows that

$$\limsup_{t-s\to\infty} \frac{1}{t-s} \ln \frac{\|X(t)\xi\|}{\|X(s)\xi\|} \leq b + \varepsilon - \alpha_1 < b + \varepsilon\,,$$

and from (1.24), it follows that

$$\liminf_{t-s\to\infty} \frac{1}{t-s} \ln \frac{\|X(t)\xi\|}{\|X(s)\xi\|} \geq a - \varepsilon + \alpha_2 > a - \varepsilon\,.$$

Since $\varepsilon > 0$ was chosen arbitrarily, it follows that

$$a \leq \liminf_{t-s\to\infty} \frac{1}{t-s} \ln \frac{\|X(t)\xi\|}{\|X(s)\xi\|} \leq \limsup_{t-s\to\infty} \frac{1}{t-s} \ln \frac{\|X(t)\xi\|}{\|X(s)\xi\|} \leq b\,.$$

Thus, $\Sigma_\xi \subset [a, b]$.

Corollary 1.3 *If the Sacker–Sell spectrum consists of points, then it coincides with the Bohl spectrum.*

Chapter 2
Linearization for Nonautonomous Differential Equations

Abstract In the first part of the chapter, we show that a hyperbolic nonautonomous differential equation can be smoothly linearized provided that the associated Sacker-Sell spectrum of the linear system satisfies a non-resonance condition. This result extends the classical Sternberg theorem to nonautonomous differential equations. The second part of the chapter is devoted to a partial linearization theorem for planar nonautonomous differential equations with one center-like direction and one hyperbolic direction. Gap conditions are formulated in terms of the dichotomy spectral intervals. The content of this chapter is a part of Cuong et al. (J Dyn Differ Equ 31(4):1279–1299, 2019) and Bonckaert et al. (J Differ Equ 258:1618–1652, 2015).

2.1 C^k Equivalence

We now define a notion of local equivalence transformation for nonautonomous differential equations with respect to a fixed and given solution. For more details, we refer the reader to [97, 99], and [114].

Definition 2.1 Consider a nonautonomous differential equation

$$\dot{x} = f(t, x), \tag{2.1}$$

with a given solution $\mu : \mathbb{R} \to \mathbb{R}^d$ and another system

$$\dot{y} = g(t, y), \tag{2.2}$$

with a solution $\widetilde{\mu} : \mathbb{R} \to \mathbb{R}^d$. If there exist $p, \widetilde{p} > 0$ and $r \in (0, p), \widetilde{r} \in (0, \widetilde{p})$ together with continuous functions

$$H : B_r(\mu) \to \mathbb{R}^d, \qquad H^{-1} : B_{\widetilde{r}}(\widetilde{\mu}) \to \mathbb{R}^d,$$

T. S. Doan, *Spectral Theory of Nonautonomous Dynamical Systems and Applications*, https://doi.org/10.1007/978-981-97-5520-2_2

where $B_r(\mu) := \{(t,x) \in \mathbb{R} \times \mathbb{R}^d : \|x - \mu(t)\| \leq r\}$ and $B_{\widetilde{r}}(\widetilde{\mu}) := \{(t,x) \in \mathbb{R} \times \mathbb{R}^d : \|x - \widetilde{\mu}(t)\| \leq \widetilde{r}\}$, then H is called a *(local)* C^k *equivalence between the system* (2.1) *with solution* μ *and system* (2.2) *with solution* $\widetilde{\mu}$, where $k \geq 1$ or $k = \infty$, if the following statements are valid:

(i) For each $t \in \mathbb{R}$ the mappings

$$H(t,\cdot) : B_r(\mu(t)) \to H(t, B_r(\mu(t))) \subset B_{\widetilde{p}}(\widetilde{\mu}(t))$$

$$H^{-1}(t,\cdot) : B_{\widetilde{r}}(\widetilde{\mu}(t)) \to H^{-1}(t, B_{\widetilde{r}}(\widetilde{\mu}(t))) \subset B_p(\mu(t))$$

are C^k diffeomorphisms (or homeomorphisms if $k = 0$) with

$$H(t, H^{-1}(t,y)) = y, \qquad \text{for } y \in B_{\widetilde{r}}(\widetilde{\mu}(t)),$$

$$H^{-1}(t, H(t,x)) = x, \qquad \text{for } x \in B_r(\mu(t)).$$

(ii) If ν is a solution of (2.1) in $B_r(\mu)$ then $H(\cdot, \nu(\cdot))$ is a solution of (2.2). If $\widetilde{\nu}$ is a solution of (2.2) in $B_{\widetilde{r}}(\widetilde{\mu})$ then $H^{-1}(\cdot, \widetilde{\nu}(\cdot))$ is a solution of (2.1).

(iii) The reference solutions are mapped uniformly onto each other:

$$\lim_{x\to 0} H(t, \mu(t) + x) = \widetilde{\mu}(t) \quad \text{uniformly in } t \in \mathbb{R},$$

$$\lim_{y\to 0} H^{-1}(t, \widetilde{\mu}(t) + y) = \mu(t) \quad \text{uniformly in } t \in \mathbb{R}.$$

Remark 2.1

(a) It is known that there is no straightforward way to extend the notion of equivalence from the autonomous setting to the nonautonomous setting. With the equivalence notion defined as in Definition 2.1, e.g., stability of the reference solution will be preserved (cp. [110, Lemma 3.6]).

(b) In principle, in the definition of C^k equivalence, there is no constraint on the smoothness of H w.r.t. the time variable t. In practice however, the easiest way to check whether solutions of one equation are mapped to solutions of the other equation is to check whether the first equation $\dot{x} = f(t,x)$ is the transformed version of the second equation $\dot{y} = g(t,y)$, after applying the transformed $y = H(t,x)$. Assuming a Carathéodory setting for $f(t,x)$, see the following definition, during the computation of $g(t,y)$ via the coordinate transform one would need $H(t,x)$ to be differentiable in t for almost all t.

Definition 2.2 (Carathéodory Function) Let $N, M \in \{1, 2, \ldots\}$, $D \subset \mathbb{R} \times \mathbb{R}^N$ open and $f : D \to \mathbb{R}^M$ a function.

(A) f is a *Carathéodory function* if for every interval $I \subset \mathbb{R}$ and every open set $U \subset \mathbb{R}^N$ with $I \times U \subset D$ the following holds:

(i) for a.a. $t \in I$ the mapping $f|_{I\times U}(t,\cdot) : U \to \mathbb{R}^M$ is continuous,
(ii) for all $x \in U$ the mapping $f|_{I\times U}(\cdot, x) : I \to \mathbb{R}^M$ is measurable with respect to the Borel σ-algebra on I and $\mathbb{R}^M$.

(B) f is a C^k *Carathéodory function*, $k \geq 0$, if

(i) for a.a. $t \in \mathbb{R}$ and all $x \in \mathbb{R}^N$ with $(t,x) \in D$ the kth partial derivative $D_x^k f(t,x)$ exists,
(ii) for every $j \in \{0,\ldots,k\}$ the mapping $D_x^j f : D \to L^j(\mathbb{R}^N;\mathbb{R}^M)$ is a Carathéodory function.

To conclude this subsection, we formulate some examples of smooth equivalences between nonautonomous systems which we then combine to get a simplified system.

Example 2.1 (Trivialization of the Reference Solution) Consider a nonautonomous differential equation $\dot{x} = f(t,x)$ on $\mathbb{R} \times \mathbb{R}^d$ with a reference solution μ. Define a transformation $H : \mathbb{R} \times \mathbb{R}^d \to \mathbb{R}^d$ by

$$H(t,x) := x - \mu(t) \qquad \text{for } (t,x) \in \mathbb{R} \times \mathbb{R}^d.$$

Then, system $\dot{x} = f(t,x)$ with a reference solution μ is C^∞ equivalent to the following system

$$\dot{y} = f(t,\mu(t)) - f(t, y + \mu(t)),$$

with the zero solution.

Example 2.2 Consider a nonautonomous linear system

$$\dot{x} = A(t)x. \tag{2.3}$$

Suppose that the dichotomy spectrum of (2.3) consists of n disjoint intervals $\lambda_n < \cdots < \lambda_1$. Then, there exists a linear transformation $S : \mathbb{R} \to \mathbb{R}^{d\times d}$ satisfying that

$$\sup_{t\in\mathbb{R}} \|S(t)\| < \infty \quad \text{and} \quad \sup_{t\in\mathbb{R}} \|S(t)^{-1}\| < \infty$$

and the C^∞ equivalence $(t,x) \mapsto S(t)x$ transforms (2.3) into a block diagonal system

$$\dot{y} = \operatorname{diag}(A_1(t),\ldots,A_n(t))$$

with locally integrable matrix functions $A_i : \mathbb{R} \to \mathbb{R}^{d_i\times d_i}$, $d_1 + d_2 + \cdots + d_n = d$ and for $i = 1,\ldots,k$ the dichotomy spectrum of the system $\dot{y}_i = A_i(t)y_i$ is λ_i. For more details, we refer the reader to [113].

Example 2.3 (Cut-Off Function) Consider a differential equation

$$\dot{x} = A(t)x + r(t, x), \tag{2.4}$$

where $r : \mathbb{R} \times B_\delta(0) \to \mathbb{R}^d$, with $\delta > 0$, is a smooth function such that $r(t, 0) = 0$ and $D_x r(t, 0) = 0$. Therefore, 0 is a solution of (2.4). Define a C^∞ cutoff function $\rho : \mathbb{R} \to [0, 1]$ by

$$\rho(a) := \begin{cases} 0, & \text{if } a \le 0; \\ \exp\left(1 - \frac{1}{a}\exp(a-1)\right), & \text{if } 0 < a < 1; \\ 1, & \text{if } \quad 1 \le a. \end{cases}$$

Using the cut-off function ρ, the function r can be extended globally as

$$\widehat{r}(t, x) := \begin{cases} \rho\left(2 - \frac{2\|x\|}{\delta}\right) r(t, x), & if \, \|x\| < \delta; \\ 0, & \text{if } \|x\| \ge \delta. \end{cases}$$

It can be easily seen that system (2.4) together with the zero solution is C^∞ equivalent to the following system

$$\dot{x} = A(t)x + \widehat{r}(t, x)$$

together with also the zero solution. Note that with this construction, the bound $\sup_{(t,x)\in\mathbb{R}\times\mathbb{R}^2} \|D_x\widehat{r}(t, x)\|$ on the derivative tends to zero when δ tends to zero provided that the second derivative $\|D_x^2\widehat{r}(t, x)\|$ is assumed to be uniformly bounded in t in a neighborhood of 0.

Combining the transformations in Examples 2.1, 2.2, and 2.3, we arrive at the following result.

Theorem 2.1 *Consider the following nonautonomous planar differential equation*

$$\dot{x} = f(t, x) \tag{2.5}$$

with a given solution $\mu : \mathbb{R} \to \mathbb{R}^d$ *which we assume to satisfy the following conditions:*

(i) $f : D \subset \mathbb{R} \times \mathbb{R}^d \to \mathbb{R}^d$ *is a smooth Carathéodory function, where* D *contains a neighborhood* $B_r(\mu)$ *for some* $r > 0$.

(ii) The linearization of (2.5) *along* μ *has bounded growth, and its dichotomy spectrum* Σ_{SS} *consists of* n *disjoint compact intervals* $\lambda_n < \cdots < \lambda_1$.

(iii) Higher order derivative terms of f *in* x *are bounded in* $B_r(\mu)$, *i.e., for all* $j \in \mathbb{N}$ *there is an* $M_j > 0$ *such that*

$$\|D_x^j f(t,x)\| \le M_j \quad \textit{for } (t,x) \in B_r(\mu).$$

The system (2.5) *together with the reference solution* μ *is locally* C^∞ *equivalent to the following system*

$$\dot{y} = \operatorname{diag}(A_1(t), \ldots, A_n(t))y + r(t,y)$$

with the zero solution, where $r : \mathbb{R} \times \mathbb{R}^d \to \mathbb{R}^d$ *is a smooth Carathéodory function satisfying the following properties:*

(i) $r(t,0) = 0$ *and* $D_y r(t,0) = 0$.
(ii) The partial derivatives of r *up to any finite order are globally bounded.*
(iii) For $i = 1, 2$, *the dichotomy spectrum of the scalar equation* $\dot{y}_i = A_i(t)y_i$ *is* $[\underline{\lambda}_i, \overline{\lambda}_i] =: \lambda_i$, *i.e., for all* $\varepsilon > 0$, *there exists* $K(\varepsilon) > 0$ *such that*

$$\frac{1}{K(\varepsilon)} e^{(\underline{\lambda}_i - \varepsilon)(t-s)} \le \Lambda_i(t,s) \le K(\varepsilon) e^{(\overline{\lambda}_i + \varepsilon)(t-s)} \qquad \textit{for } t \ge s, \tag{2.6}$$

where $\Lambda_i(t,s)$ *denotes the evolution operator of the equation* $\dot{y}_i = A_i(t)y_i$.

2.2 Smooth Linearization

2.2.1 Motivation and Sternberg's Theorem

The Hartman-Grobman theorem for nonautonomous differential equations [97] states that if f is a C^0-small map from $\mathbb{R} \times \mathbb{R}^d$ into $\mathbb{R}^d$ and also small Lipschitz with respect to the second argument, then there is a continuous family of homeomorphisms $H(t,\cdot)$, $t \in \mathbb{R}$, of $\mathbb{R}^d$, with $\sup_{(t,x)\in\mathbb{R}\times\mathbb{R}^d} \|H(t,x) - x\| < \infty$ and sending solutions $t \mapsto \mu(t)$ of the differential equation

$$\dot{x} = A(t)x + f(t,x), \tag{2.7}$$

onto solutions $t \mapsto H(t,\mu(t))$ of the linear differential equation

$$\dot{x} = A(t)x, \tag{2.8}$$

and vice versa with its inverse $H^{-1}(t,\cdot)$ with respect to the second argument. In this section, we show that $H(t,\cdot)$, $t \in \mathbb{R}$, can be chosen as a C^ℓ diffeomorphism for any given $\ell \in \mathbb{N}$, provided the linear differential equation (2.8) satisfies a non-resonance condition which is formulated in terms of the Sacker-Sell spectrum, the larger the ℓ the more non-resonance conditions have to be satisfied (Theorem 2.2). This result generalizes the classical Sternberg theorem [35, 116, 117] to nonautonomous differential equations.

More precisely, for $k \in \mathbb{N}$, we consider C^k Carathéodory differential equations of the form (2.7), with a locally integrable matrix function $A : \mathbb{R} \to \mathbb{R}^{d\times d}$ and a C^k Carathéodory function $f : \mathbb{R}\times\mathbb{R}^d \to \mathbb{R}^d$, i.e., for a.a. $t \in \mathbb{R}$ $f(t,\cdot)$ is continuous, for all $x \in \mathbb{R}^d$ $f(\cdot, x)$ is measurable, for a.a. $t \in \mathbb{R}$ and all $x \in \mathbb{R}^d$ the partial derivative $D_x^k f(t,x)$ exists and $D_x^j f$ is a Carathéodory function for every $j \in \{1,\dots,k\}$; see Definition 2.2 . In order to linearize (2.7) smoothly along its trivial solution, we assume the following conditions:

(A1) Equation (2.7) is Taylor expanded along its trivial solution:

$$f(t,0) = 0 \quad \text{and} \quad D_x f(t,0) = 0 \qquad \text{for a.a. } t \in \mathbb{R}.$$

(A2) Linearity: (2.8) has bounded growth [47], i.e., with the evolution operator $\Phi\colon \mathbb{R}\times\mathbb{R} \to \mathbb{R}^{d\times d}$ of (2.8) there exist $K, a > 0$ such that

$$\|\Phi(t,s)\| \le K e^{a|t-s|} \qquad \text{for all } t, s \in \mathbb{R}.$$

(A3) Nonlinearity: There exists $M > 0$ such that

$$\|D_x^j f(t,x)\| \le M \qquad \text{for } j = 0,\dots,k, \text{ a.a. } t \in \mathbb{R} \text{ and all } x \in \mathbb{R}^d.$$

The global boundedness condition (A3) can be achieved by cutting of outside of a tubular neighborhood $\mathbb{R}\times B_r(0)$, $r > 0$, of the trivial solution or can be replaced by a local condition (see Example 2.3 and Theorem 2.1).

For $k \in \mathbb{N}$ and a locally integrable matrix function $A : \mathbb{R} \to \mathbb{R}^{d\times d}$ satisfying (A2), we define the set of admissible differential equations

$$\mathcal{O}^k(A) := \{\dot{x} = A(t)x + f(t,x)\colon f \text{ is a } C^k \text{ Carathéodory function which satisfies (A1) and (A3)}\}.$$

If (2.8) is hyperbolic, i.e., if $0 \notin \Sigma_{SS}(A)$, then all systems in $\mathcal{O}^1(A)$ are topologically equivalent (C^0 equivalent) to (2.8); see [97]. Our main result is that if (2.8) is hyperbolic, then for any given $\ell \in \mathbb{N}$, there exists a $k \in \mathbb{N}$ (which depends on ℓ and the Sacker Sell spectrum $\Sigma_{SS}(A)$) such that the following implication holds: $\langle$ (2.8) is nonresonant up to order k $\rangle$ $\Rightarrow$ $\langle$ all systems in $\mathcal{O}^{k+2}(A)$ are C^ℓ equivalent to (2.8) $\rangle$.

Theorem 2.2 (Sternberg Theorem for Nonautonomous Differential Equations) *Let $\ell \in \mathbb{N}$, and let $\Sigma_{SS}(A) = \lambda_1 \cup \lambda_2 \cup \cdots \cup \lambda_n$ denote the Sacker Sell spectrum of the linear system* (2.8). *Assume that* $0 \notin \Sigma_{SS}(A)$. *Then with*

$$\Sigma_s := \Sigma_{SS}(A) \cap \mathbb{R}_{<0} \qquad \textit{and} \qquad \Sigma_u := \Sigma_{SS}(A) \cap \mathbb{R}_{>0}$$

there exists a minimal $k \in \mathbb{N}$, $k \geq \ell$, *such that*

$$(k-\ell)\Sigma_u > \Sigma_u - \ell\Sigma_s \qquad \textit{and} \qquad (k-\ell)\Sigma_s < \Sigma_s - \ell\Sigma_u. \tag{2.9}$$

If the non-resonance condition

$$\lambda_j \cap \sum_{i=1}^n k_i\lambda_i = \emptyset \qquad \textit{for all } (k_1,\ldots,k_n) \in \mathbb{N}_0^n \textit{ with } 2 \leq \sum_{i=1}^n k_i \leq k$$

holds, then all systems in $\mathcal{O}^{k+2}(A)$ *are* C^ℓ *equivalent to their linearization* (2.8).

To prove the preceding theorem, we need several preparatory results. First, in Sect. 2.2.2, we establish a result on flatting invariant manifolds and eliminating nonresonant Taylor terms for systems in $\mathcal{O}^{k+2}(A)$. Section 2.2.3 is devoted to studying discrete-time systems associated with systems in $\mathcal{O}^{k+2}(A)$. Using preparatory results in Sects. 2.2.2 and 2.2.3, the Sternberg theorem for nonautonomous differential equations (Theorem 2.2) is proved in Sect. 2.2.4. An important technique in the proof of the nonautonomous Sternberg theorem, namely, method of path, is established in Sect. 2.2.5.

2.2.2 Flatting Stable/Unstable Manifolds and Eliminating Nonresonant Taylor Terms

In this subsection, we cite a short version of smooth invariant manifold theory for nonautonomous differential equations from [110, Satz 4.32]. A discrete-time version of this theorem is given in [15] and [104].

Let $A\colon \mathbb{R} \to \mathbb{R}^{d\times d}$ be locally integrable with $0 \notin \Sigma_{SS}(A)$. By Example 2.2, there exists a kinematic similarity $S\colon \mathbb{R} \to \mathbb{R}^{d\times d}$ such that $y(t) = S(t)x(t)$ transforms (2.8) into a block diagonal system

$$\dot{y} = \operatorname{diag}(A_1(t),\ldots,A_n(t))y.$$

Since

$$H\colon \mathbb{R}\times\mathbb{R}^d \to \mathbb{R}^d, \quad (t,x) \mapsto H(t,x) := S(t)x,$$

satisfies conditions (i), (ii), and (iii) of Definition 2.1 with inverse $H^{-1}(t,\cdot) = S(t)^{-1}$, we can assume w.l.o.g. that

$$A(t) = \operatorname{diag}(A_1(t),\ldots,A_n(t))$$

with $\Sigma_{SS}(A_i) = \lambda_i$ for $i = 1, \dots, n$, $\Sigma_{SS}(A) = \cup_{i=1}^{n} \lambda_i$ and $\lambda_1 > \lambda_2 > \cdots > \lambda_n$. With the unique $m \in \{0, \dots, n\}$ such that

$$\lambda_1 > \cdots > \lambda_m > 0 > \lambda_{m+1} > \cdots > \lambda_n,$$

define

$$\begin{aligned} A^u(t) &:= \operatorname{diag}(A_1(t), \dots, A_m(t)) \in \mathbb{R}^{d_u \times d_u}, \\ A^s(t) &:= \operatorname{diag}(A_{m+1}(t), \dots, A_n(t)) \in \mathbb{R}^{d_s \times d_s}, \end{aligned} \tag{2.10}$$

where $d_u + d_s = d$, and if $m = 0$ then $d_u = 0$, $A^s = A$, if $m = d$ then $d_s = 0$, $A^u = A$. Let π_u and π_s denote the canonical projections of $\mathbb{R}^d = \mathbb{R}^{d_u} \times \mathbb{R}^{d_s}$ into $\mathbb{R}^{d_u}$ and $\mathbb{R}^{d_s}$, respectively. Rewriting a differential equation (2.7) in $\mathcal{O}^k(A)$ in (x^u, x^s)-coordinates, with $x^u := \pi_u x$, $x^s := \pi_s x$, gives

$$\begin{aligned} \dot{x}^u &= A^u(t)x^u + \pi_u f(t, x^u, x^s), \\ \dot{x}^s &= A^s(t)x^s + \pi_s f(t, x^u, x^s). \end{aligned} \tag{2.11}$$

Let $\varphi(\cdot, t_0, x_0^u, x_0^s)$ denote the solution of the initial value problem (2.11), $x(t_0) = (x_0^u, x_0^s)$. Then nonautonomous invariant manifold theory (also called integral manifold theory) yields *unstable* and *stable manifolds*

$$\begin{aligned} \mathcal{U} &= \left\{(t, x^u, x^s) : \lim_{s \to -\infty} \varphi(s, t, x^u, x^s) = 0\right\} \\ \mathcal{S} &= \left\{(t, x^u, x^s) : \lim_{s \to \infty} \varphi(s, t, x^u, x^s) = 0\right\} \end{aligned}$$

of the zero solution of (2.11), respectively (see, e.g., [13] or [110]). We say that $\mathcal{U}$ and $\mathcal{S}$ are *flat*, if they coincide with the unstable and stable manifolds of the linearization

$$\begin{aligned} \dot{x}^u &= A^u(t)x^u, \\ \dot{x}^s &= A^s(t)x^s, \end{aligned}$$

which are the subspaces (or, more precisely, the trivial vector bundles over the base space $\mathbb{R}$) $\mathbb{R} \times \mathbb{R}^{d_u} \times \{0\} \subseteq \mathbb{R}^{1+d}$ and $\mathbb{R} \times \{0\} \times \mathbb{R}^{d_s} \subseteq \mathbb{R}^{1+d}$, respectively. Using the invariance of $\mathcal{U}$ and $\mathcal{S}$, one can characterize flatness by

$$\begin{aligned} \mathcal{U} = \mathbb{R} \times \mathbb{R}^{d_u} \times \{0\} &\Leftrightarrow \pi_s f(t, x^u, 0) = 0 \text{ for all } (t, x^u) \in \mathbb{R} \times \mathbb{R}^{d_u}, \\ \mathcal{S} = \mathbb{R} \times \{0\} \times \mathbb{R}^{d_u} &\Leftrightarrow \pi_u f(t, 0, x^s) = 0 \text{ for all } (t, x^s) \in \mathbb{R} \times \mathbb{R}^{d_s}. \end{aligned}$$

In what follows, we construct a C^k equivalence between an $\mathcal{O}^{k+2}(A)$ system (2.11) and a system with flat stable and unstable manifolds. Moreover, using [114], we will also eliminate the higher-order Taylor terms $D_x^i f(t,0)$ for $i = 2,\dots,k$ with a C^k equivalence, provided $\Sigma_{SS}(A)$ satisfies certain non-resonance conditions, in order to arrive at a system which additionally satisfies the following two conditions.

(A4) Stable and unstable manifolds are flat: $\pi_u f(t,0,x^s) = 0$, $\pi_s f(t,x^u,0) = 0$ for a.a. $t \in \mathbb{R}$ and all $(x^u,x^s) \in \mathbb{R}^{d_u} \times \mathbb{R}^{d_s}$,

(A5) Taylor terms up to order k vanish: $D_x^i f(t,0) = 0$ for $i = 2,\dots,k$ and a.a. $t \in \mathbb{R}$. (Recall that $D_x f(t,0) = 0$ by (A1).)

Define $\mathcal{O}^k_{\text{flat}}(A) \subseteq \mathcal{O}^k(A)$ by

$$\mathcal{O}^k_{\text{flat}}(A) := \mathcal{O}^k(A) \cap \{\dot{x} = A(t)x + f(t,x) : f \text{ satisfies (A4) and (A5)}\}.$$

In the following proposition, we show that under the non-resonance condition of Theorem 2.2, any system in $\mathcal{O}^{k+2}(A)$ is C^k equivalent to a system in $\mathcal{O}^k(A)_{\text{flat}}$.

Proposition 2.1 (Flattening of Invariant Manifolds and Elimination of Nonresonant Terms) *Let $\Sigma_{SS}(A) = \lambda_1 \cup \lambda_2 \cup \dots \cup \lambda_n$ denote the Sacker Sell spectrum of the linear system* (2.8). *Assume that $0 \notin \Sigma_{SS}(A)$. Let $k \in \mathbb{N}$. If the non-resonance condition*

$$\lambda_j \cap \sum_{i=1}^n k_i \lambda_i = \emptyset \qquad \textit{for all } (k_1,\dots,k_n) \in \mathbb{N}_0^n \textit{ with } 2 \le \sum_{i=1}^n k_i \le k \tag{2.12}$$

holds, then any system in $\mathcal{O}^{k+2}(A)$ is C^k equivalent to a system in $\mathcal{O}^k_{\text{flat}}(A)$.

Proof Consider an arbitrary system

$$\dot{x} = A(t)x + f(t,x) \tag{2.13}$$

in $\mathcal{O}^{k+2}(A)$ with solution $\varphi(\cdot,t_0,x_0^u,x_0^s)$ satisfying $\varphi(t_0,t_0,x_0^u,x_0^s) = (x_0^u,x_0^s)$. By definition (2.10) of A^u and A^s, we have $\Sigma_{SS}(A^u) > 0 > \Sigma_{SS}(A^s)$. Let Φ^u and Φ^s denote the evolution operators of $\dot{x}^u = A^u(t)x^u$ and $\dot{x}^s = A^s(t)x^s$, respectively. Then by spectral theorem of Sacker-Sell spectrum on the whole real line (Theorem 1.2), there exist $K,\alpha > 0$ such that

$$\|\Phi^u(t,s)\| \le Ke^{\alpha(t-s)} \qquad \text{for } t \le s,$$
$$\|\Phi^s(t,s)\| \le Ke^{-\alpha(t-s)} \qquad \text{for } t \ge s.$$

Step 1 (Flattening of stable and unstable manifolds): Using [13, Theorem 4.1], there exists an $L = L(K,\alpha) > 0$ such that if $\sup_{t\in\mathbb{R},x\in\mathbb{R}^d} \|D_x f(t,x)\| \le L$, then there exist stable and unstable manifolds of the trivial solution of (2.13). Using assumption (A3) and the mean value theorem, it follows that

$$\sup_{t\in\mathbb{R},x\in\mathbb{R}^d} \|D_x f(t,x)\| \le \sup_{s\in\mathbb{R},y\in\mathbb{R}^d} \|D_x^2 f(s,y)\| \|x\| \le M\|x\|$$

and replacing f by a cut-off $\tilde{f}$ (see, e.g., Example 2.3) which agrees with f on a tubular neighborhood $\mathbb{R}\times B_r(0)\subset\mathbb{R}\times\mathbb{R}^d$ for some sufficiently small $r>0$ and still satisfies (A3), we get $\sup_{t\in\mathbb{R},x\in\mathbb{R}^d}\|D_x\tilde{f}(t,x)\|\le L$. The cut-off procedure is a C^k equivalence (see Example 2.3), and for ease of notation, we omit the tilde and write again f instead of $\tilde{f}$.

By Aulbach and Wanner [13, Theorem 4.1], there exists a continuous map $\mathfrak{u}\colon\mathbb{R}\times\mathbb{R}^{d_u}\to\mathbb{R}^{d_s}$ whose graph

$$\mathcal{U} := \left\{(t,x^u,\mathfrak{u}(t,x^u))\colon (t,x^u)\in\mathbb{R}\times\mathbb{R}^{d_u}\right\}$$

can be characterized dynamically as

$$\mathcal{U} = \left\{(t,x^u,x^s): \lim_{s\to-\infty}\varphi(s,t,x^u,x^s)=0\right\}.$$

$\mathcal{U}$ is the unstable manifold of the trivial solution of (2.13), $\mathfrak{u}(\cdot,0)=0$, and the invariance equation

$$\pi_s\circ\varphi(t,s,x^u,\mathfrak{u}(s,x^u)) = \mathfrak{u}\big(t,\pi_u\circ\varphi(t,s,x^u,\mathfrak{u}(s,x^u))\big) \tag{2.14}$$

holds for $t,s\in\mathbb{R}$, $(x^u,x^s)\in\mathbb{R}^d$. By Siegmund [110, Theorem 5.20] $\mathfrak{u}$ is a C^{k+2} Carathéodory function with globally bounded derivatives and $\frac{\partial\mathfrak{u}}{\partial x^u}(t,0)=0$ for $t\in\mathbb{R}$. Define $H\colon\mathbb{R}\times\mathbb{R}^d\to\mathbb{R}^d$ by

$$H(t,x^u,x^s) := (x^u, x^s-\mathfrak{u}(t,x^u)).$$

Then $H(\cdot,0)=0$, $H(t,\cdot)$ are C^{k+2} diffeomorphisms with inverse $H^{-1}(t,x)=(x^u,x^s+\mathfrak{u}(t,x^u))$ and $\lim_{x\to0}H(t,x)=\lim_{x\to0}H^{-1}(t,x)=0$ uniformly in $t\in\mathbb{R}$ due to the fact that $\sup_{t\in\mathbb{R},x^u\in\mathbb{R}^{d_u}}\|\frac{\partial\mathfrak{u}}{\partial x^u}(t,x^u)\|<\infty$.

Taking derivatives of both sides of the invariance equation (2.14) yields

$$\begin{aligned}\frac{\partial\mathfrak{u}}{\partial t}(t,x^u) &= A^s(t)\mathfrak{u}(t,x^u)+\pi_s f(t,x^u,\mathfrak{u}(t,x^u))\\ &\quad -\frac{\partial\mathfrak{u}}{\partial x^u}(t,x^u)\big(A^u(t)x^u+\pi_u f(t,x^u,\mathfrak{u}(t,x^u))\big)\end{aligned}$$

which implies that H transforms (2.13) to

$$\dot{x} = A(t)x+g(t,x) \tag{2.15}$$

with

$$\pi_u g(t, x^u, x^s) = \pi_u f(t, x^u, x^s + \mathfrak{u}(t, x^u)),$$

$$\pi_s g(t, x^u, x^s) = \pi_s f(t, x^u, x^s + \mathfrak{u}(t, x^u)) - \pi_s f(t, x^u, \mathfrak{u}(t, x^u))$$
$$- \frac{\partial \mathfrak{u}}{\partial x^u}(t, x^u)\big(\pi_s f(t, x^u, x^s + \mathfrak{u}(t, x^u)) - \pi_s f(t, x^u, \mathfrak{u}(t, x^u))\big).$$

System (2.15) is in $\mathcal{O}^{k+1}(A)$; it satisfies additionally that $\pi_s g(t, x^u, 0) = 0$, i.e., its unstable manifold is flat, and (2.13) is C^{k+1} equivalent to (2.15). Similarly, one can flatten the stable manifold $\mathcal{S} := \{(t, \mathfrak{s}(t, x^s), x^s) : (t, x^s) \in \mathbb{R} \times \mathbb{R}^{d_s}\}$ of (2.15) with a C^k equivalence $(t, x^u, x^s) \mapsto (x^u - \mathfrak{s}(t, x^s), x^s)$. By composing both transformations, (2.13) is C^k equivalent to a system in $\mathcal{O}^k(A)$ which satisfies condition (A4), i.e., the stable and unstable manifolds are flat.
Step 2 (Elimination of nonresonant terms): If $k = 1$, no terms need to be eliminated. If $k \geq 2$, the normal form theorem for nonautonomous differential equations [114] implies that all nonresonant Taylor terms up to order k can be eliminated by a C^k equivalence. By assumption (2.12), all Taylor terms up to order k are nonresonant, and therefore the simplified system from Step 1 is C^k equivalent to a system in $\mathcal{O}^k_{\text{flat}}(A)$.

2.2.3 *Smooth Equivalence of Discrete Time Systems*

Nonautonomous difference equations (also called discrete time systems) of the form

$$x_{n+1} = B_n x_n + g_n(x_n) \tag{2.16}$$

with a matrix-valued function $B : \mathbb{Z} \to \mathbb{R}^{d\times d}$, $n \mapsto B_n$, and a function $g : \mathbb{Z}\times\mathbb{R}^d \to \mathbb{R}^d$, $(n, x) \mapsto g_n(x)$, such that g_n is C^k, arise, e.g., as time-1 maps of (2.7) by setting $B_n := \Phi(n + 1, n)$ and $g_n(x) := \varphi(n + 1, n, x) - \Phi(n + 1, n)x$ with the evolution operator Φ of (2.8) and the solution φ of (2.7). Then obviously every solution $n \mapsto x_n$ of (2.16) satisfies $x_n = \varphi(n, m, x_m)$.

For discrete time systems (2.16) with $g_n(0) = 0$ and $g_n'(0) = 0$ $(n \in \mathbb{Z})$, a Hartman-Grobman topological linearization result and a smooth Poincaré normal form theory are already established (see, e.g., [14] and the references therein, and [115], respectively). To the best of our knowledge, a Sternberg linearization theorem for (2.16) was not proven yet. For an autonomous version of Sternberg's linearization theorem for difference equations, see [32] and [94] and the references therein.

In this section, we utilize a result on the "continuous dependence of the smooth change of coordinates in parametrized normal form theorems" for autonomous difference equations [29] which we extend to the nonautonomous discrete time case (2.16) and then relate to the construction of smooth linearizations of (2.7). According to Proposition 2.1, in order to prove Theorem 2.2, it remains to show that all nonresonant systems in $\mathcal{O}^k_{\text{flat}}(A)$ are C^r equivalent to their linearization (2.8). As a main step in the proof, we discuss discrete time systems (2.16) which arise as time-κ maps ($\kappa > 0$ large) of systems (2.7) in $\mathcal{O}^k_{\text{flat}}(A)$.

2.2.3.1 Associated Discrete-Time Systems

Definition 2.3 (Associated Time-κ System) For $k \in \mathbb{N}$ and a locally integrable matrix function $A : \mathbb{R} \to \mathbb{R}^{d\times d}$ satisfying (A2), consider a differential equation

$$\dot{x} = A(t)x + f(t,x) \tag{2.17}$$

in $\mathcal{O}^k(A)$ with the solution $\varphi(\cdot, t_0, x_0)$ of (2.17), $x(t_0) = x_0$. Let $\kappa \in \mathbb{R}$, $\kappa > 0$. For $n \in \mathbb{Z}$, $x \in \mathbb{R}^d$, define

$$F_n^{(\kappa)}(x) := \varphi((n+1)\kappa, n\kappa, x) \quad \text{and}$$
$$A_n^{(\kappa)} := \Phi((n+1)\kappa, n\kappa), \quad f_n^{(\kappa)}(x) := F_n^{(\kappa)}(x) - A_n^{(\kappa)}x$$

with the evolution operator Φ of (2.8). Then

$$x_{n+1} = F_n^{(\kappa)}(x_n) \quad \text{and equivalently} \quad x_{n+1} = A_n^{(\kappa)}x_n + f_n^{(\kappa)}(x_n) \tag{2.18}$$

are called the *associated time-κ system* of (2.17).

For $k \in \mathbb{N}$ and a C^k map $f : \mathbb{R}^d \to \mathbb{R}^d$, let $\|f\|_k := \max_{0\le i\le k} \sup_{x\in\mathbb{R}^d} \|D^i f(x)\|$. Define

$$C^k(\mathbb{R}^d) := \{f : \mathbb{R}^d \to \mathbb{R}^d : f \text{ is } C^k \text{ with } \|f\|_k < \infty\},$$
$$C^k_1(\mathbb{R}^d) := \{f \in C^k(\mathbb{R}^d) : f(0) = 0, Df(0) = 0\},$$
$$C^k_{\text{flat}}(\mathbb{R}^d) := \{f \in C^k(\mathbb{R}^d) : D^i f(0) = 0 \text{ for } i = 0, \ldots, k\}.$$

Lemma 2.1 *For an associated time-κ system* (2.18) *of* (2.17), *the following statements hold:*

(a) $\sup_{n\in\mathbb{Z}} \|A_n^{(\kappa)}\| < \infty$ *and* $\sup_{n\in\mathbb{Z}} \|(A_n^{(\kappa)})^{-1}\| < \infty$.
(b) If (2.17) *is in* $\mathcal{O}^k_{\text{flat}}(A)$, *then* $f_n^{(\kappa)} \in C^k_{\text{flat}}(\mathbb{R}^d)$ *for* $n \in \mathbb{N}$.
(c) If (2.17) *is in* $\mathcal{O}^k_{\text{flat}}(A)$, *then* $\pi_u f_n^{(\kappa)}(0, x^s) = 0$ *and* $\pi_s f_n^{(\kappa)}(x^u, 0) = 0$ *for* $n \in \mathbb{Z}$ *and* $(x^u, x^s) \in \mathbb{R}^d$.

Proof

(a) By (A2), $\|A_n^{(\kappa)}\| = \|\Phi((\kappa+1)n, \kappa n)\| \le Ke^{a\kappa}$ and similarly for $\|(A_n^{(\kappa)})^{-1}\|$.
(b) Using $f_n^{(\kappa)}(x) = \varphi((n+1)\kappa, n\kappa, x) - \Phi((n+1)\kappa, n\kappa)x$, this follows for $i = 0$ from $\varphi(\cdot, \cdot, 0) = 0$, for $i = 1$ from $\frac{\partial\varphi}{\partial x}((n+1)\kappa, n\kappa, 0) = \Phi((n+1)\kappa, n\kappa)$ and

for $i = 2, \dots k$ from Taylor expanding the C^k function $x \mapsto \varphi((n+1)\kappa, n\kappa, x)$ in $x = 0$.

(c) Using $f_n^{(\kappa)}(x) = \varphi((n+1)\kappa, n\kappa, x) - \Phi((n+1)\kappa, n\kappa)x$, the claim follows from the invariance equation (2.14) for stable and unstable manifolds of (2.17) which are flat by (A4), i.e., $\pi_u f(t, 0, x^s) = 0$ and $\pi_s f(t, x^u, 0) = 0$ for a.a. $t \in \mathbb{R}$ and all $(x^u, x^s) \in \mathbb{R}^{d_u} \times \mathbb{R}^{d_s}$, or equivalently, $\mathfrak{u}(\cdot, 0) = 0$ and $\mathfrak{s}(\cdot, 0) = 0$.

2.2.3.2 Discrete Time Characterization of C^k Equivalence

Theorem 2.3 (Time-κ Characterization of C^k Equivalence) *Two systems*

$$\dot{x} = A(t)x + f(t, x) \tag{2.19}$$

and

$$\dot{y} = A(t)y + g(t, y) \tag{2.20}$$

in $\mathcal{O}^k(A)$ are C^k equivalent, if there exists a $\kappa > 0$ such that their associated time-κ systems

$$x_{n+1} = F_n^{(\kappa)}(x_n)$$

and

$$y_{n+1} = G_n^{(\kappa)}(y_n)$$

are C^k equivalent, i.e., if there exist $p, \widetilde{p} > 0$ and $r \in (0, p), \widetilde{r} \in (0, \widetilde{p})$ and maps

$$T_n \colon B_r(0) \to \mathbb{R}^d \qquad \text{and} \qquad S_n \colon B_{\widetilde{r}}(0) \to \mathbb{R}^d,$$

such that the following statements hold:

1. *For each $n \in \mathbb{Z}$ the maps*

$$T_n \colon B_r(0) \to T_n(B_r(0)) \subset B_{\widetilde{p}}(0),$$
$$S_n \colon B_{\widetilde{r}}(0) \to S_n(B_{\widetilde{r}}(0)) \subset B_p(0),$$

are C^k diffeomorphisms with

$$T_n(S_n(y)) = y \quad \text{for all } y \in B_{\widetilde{r}}(0) \cap S_n^{-1}(B_r(0)),$$
$$S_n(T_n(x)) = x \quad \text{for all } x \in B_r(0) \cap T_n^{-1}(B_{\widetilde{r}}(0)).$$

2. *For $x \in B_r(0)$, $y \in B_{\widetilde{r}}(0)$ with $F_n^{(\kappa)}(x) \in B_r(0)$, $G_n^{(\kappa)}(y) \in B_{\widetilde{r}}(0)$*

$$T_{n+1}(F_n^{(\kappa)}(x)) = G_n^{(\kappa)}(T_n(x)) \quad \text{and} \quad S_{n+1}(G_n^{(\kappa)}(y)) = F_n^{(\kappa)}(S_n(y)).$$

3. The trivial solutions are mapped uniformly onto each other:

$$\lim_{x\to 0} T_n(x) = 0 \quad \text{and} \quad \lim_{y\to 0} S_n(y) = 0 \qquad \text{uniformly in } n \in \mathbb{Z}.$$

Proof With the solutions φ and ψ of (2.19) and (2.20), respectively,

$$F_n^{(\kappa)}(x) = \varphi((n+1)\kappa, n\kappa, x) \quad \text{and} \quad G_n^{(\kappa)}(y) = \psi((n+1)\kappa, n\kappa, y).$$

Since (2.19) and (2.20) are in $\mathcal{O}^k(A)$, the functions f and g satisfy the global boundedness condition (A3). By Gronwall's inequality [71, Chapter 3, Theorem 1.1], there exist $r^* > 0$ and $\tilde{r}^* > 0$ such that for all $n \in \mathbb{Z}$

$$\begin{aligned} &\varphi(\kappa n, t, x) \in B_r(0) && \text{for } t \in [\kappa n, \kappa(n+1)], x \in B_{r^*}(0), \\ &\psi(\kappa n, t, y) \in B_{\tilde{r}}(0) && \text{for } t \in [\kappa n, \kappa(n+1)], y \in B_{\tilde{r}^*}(0). \end{aligned} \tag{2.21}$$

By assumption (iii), and by shrinking r^* and $\tilde{r}^*$ if necessary, we can assume additionally that

$$\begin{aligned} &T_n(\varphi(\kappa n, t, x)) \in B_r(0) && \text{for } t \in [\kappa n, \kappa(n+1)], x \in B_{r^*}(0), \\ &S_n(\psi(\kappa n, t, y)) \in B_{\tilde{r}}(0) && \text{for } t \in [\kappa n, \kappa(n+1)], y \in B_{\tilde{r}^*}(0). \end{aligned} \tag{2.22}$$

Define $H : \mathbb{R} \times B_{r^*}(0) \to \mathbb{R}^d$ and $H^{-1} : \mathbb{R} \times B_{\tilde{r}^*}(0) \to \mathbb{R}^d$ by

$$\begin{aligned} H(t, x) &:= \psi\big(t, \kappa n, T_n(\varphi(\kappa n, t, x))\big) && \text{for } x \in B_{r^*}(0), \\ H^{-1}(t, y) &:= \varphi\big(t, \kappa n, S_n(\psi(\kappa n, t, y))\big) && \text{for } y \in B_{\tilde{r}^*}(0), \end{aligned} \tag{2.23}$$

for $t \in [\kappa n, \kappa(n+1))$. By (2.21) and (2.22), the functions H and H^{-1} are well-defined, and to conclude the proof, we show that H is a C^k equivalence between (2.19) and (2.20) by verifying that H and H^{-1} satisfy conditions (i), (ii), and (iii) in Definition 2.1.

Step 1 (Verification of Definition 2.1(i)): The functions $H(t, \cdot)$ and $H^{-1}(t, \cdot)$ are C^k, since T_n and S_n are C^k diffeomorphisms and the solutions φ and ψ are C^k by the classical result on the smooth dependence of solutions of ordinary differential equations on the initial value; see, e.g., [71, Chapter V, Theorem 4.1]. Now let $y \in B_{\tilde{r}^*}(0)$ satisfy that $H^{-1}(t, y) \in B_{r^*}(0)$. By definition (2.23) of H and H^{-1}, we have

$$H(t, H^{-1}(t, y)) = \psi\big(t, \kappa n, T_n \circ S_n(\psi(\kappa n, t, y))\big).$$

By (2.21), (2.22) and assumption (i), $T_n \circ S_n(\psi(\kappa n, t, y)) = \psi(\kappa n, t, y)$. Therefore, $H(t, H^{-1}(t, y)) = y$. Similarly, for $x \in B_{r^*}(0)$ with $H(t, x) \in B_{\tilde{r}^*}(0)$, we also have $H^{-1}(t, H(t, x)) = x$.

Step 2 (Verification of Definition 2.1(ii)): Let $\mu(t)$ be an arbitrary solution of (2.19) in $B_{r^*}(0)$. By definition (2.23) of H, for $n \in \mathbb{Z}$ and $t \in [\kappa n, \kappa(n+1))$

$$\begin{aligned} H(t, \mu(t)) &= \psi\big(t, \kappa n, T_n(\varphi(\kappa n, t, \mu(t)))\big) \\ &= \psi(t, \kappa n, T_n(\mu(\kappa_n))), \end{aligned}$$

where we use the fact that $\mu(t)$ is a solution of (2.7) to obtain the last equality. Hence, $H(t, \mu(t))$ is a solution of (2.20) on $[\kappa n, \kappa(n+1))$ and, by continuity, also on $\mathbb{R}$. Similarly, $H^{-1}(t, \nu(t))$ is a solution of (2.19) for any solution $\nu(t)$ of (2.20) in $B_{\tilde{r}^*}(0)$.

Step 3 (Verification of Definition 2.1(iii)): Since (2.19) and (2.20) are in $\mathcal{O}^k(A)$, the functions f and g satisfy the global boundedness condition (A3) with bound $M > 0$ and therefore by the mean value theorem

$$\|f(t, x)\|, \|g(t, x)\| \le M\|x\| \qquad \text{for a.a. } t \in \mathbb{R}, x \in \mathbb{R}^d.$$

Using the variation of constants formula, we obtain for all $t \in [\kappa n, \kappa(n+1)]$

$$\psi(t, \kappa n, x) = \Phi(t, \kappa n)x + \int_{\kappa n}^{t} \Phi(t, s) f(s, \psi(s, \kappa n, x))\, ds,$$

with the evolution operator Φ of $\dot{x} = A(t)x$, which together with (A2) implies that

$$\|\psi(t, \kappa n, x)\| \le K e^{a\kappa}\|x\| + MKe^{a\kappa} \int_{\kappa n}^{t} \|\psi(s, \kappa n, x)\|\, ds.$$

Then, using Gronwall's inequality [71, Chapter 3, Theorem 1.1], there exists $C_1 > 0$ such that

$$\|\psi(t, \kappa n, x)\| \le C_1\|x\| \qquad \text{for } t \in [\kappa n, \kappa(n+1)).$$

Similarly, there exists $C_2 > 0$ such that

$$\|\varphi(\kappa n, t, x)\| \le C_2\|x\| \qquad \text{for } t \in [\kappa n, \kappa(n+1)).$$

As a consequence, by (2.23), we get for $t \in [\kappa n, \kappa(n+1))$

$$\|H(t, x)\| \le C_1 \sup_{z \in B_{C_2\|x\|}(0)} T_n(z)$$

and by taking the supremum over $t \in \mathbb{R}$, we have

$$\sup_{t\in\mathbb{R}} \|H(t,x)\| \le C_1 \sup_{n\in\mathbb{Z}} \sup_{z\in B_{C_2\|x\|}(0)} T_n(z).$$

Assumption (iii) implies that $\lim_{x\to 0} H(t,x) = 0$ uniformly in $t \in \mathbb{R}$. Similarly, we also have $\lim_{y\to 0} H^{-1}(t,y) = 0$ uniformly in $t \in \mathbb{R}$.

2.2.3.3 Associated Discrete Time Systems with Spectral Gap and Flat Nonlinearity

Recall that $A\colon \mathbb{R} \to \mathbb{R}^{d\times d}$ is of the form (2.10), i.e., $A(t) = \operatorname{diag}(A^u(t), A^s(t))$ with $0 \notin \Sigma_{\mathrm{SS}}(A) = \Sigma^u \cup \Sigma^s$, $\Sigma^u = \Sigma_{\mathrm{SS}}(A) \cap \mathbb{R}_{>0}$, $\Sigma^s = \Sigma_{\mathrm{SS}}(A) \cap \mathbb{R}_{<0}$ and corresponding coordinates $x = (x^u, x^s) \in \mathbb{R}^{d_u} \times \mathbb{R}^{d_s}$. In this subsection, we discuss discrete-time systems

$$x_{n+1} = B_n x_n + g_n(x_n), \qquad (n \in \mathbb{Z})$$

with $B_n = \operatorname{diag}(B_n^u, B_n^s) \in \mathrm{Gl}(d,\mathbb{R})$, $B^u : \mathbb{Z} \to \mathbb{R}^{d_u\times d_u}$, $B^s : \mathbb{Z} \to \mathbb{R}^{d_s\times d_s}$, $g : \mathbb{Z} \times \mathbb{R}^n \to \mathbb{R}^n$, which satisfy the following conditions for fixed $k, \ell \in \mathbb{N}$:

(B1) Linearity: B is bounded, i.e., there exists $M > 0$ such that

$$\sup_{n\in\mathbb{Z}} \|B_n\| \le M \quad \text{and} \quad \sup_{n\in\mathbb{Z}} \|B_n^{-1}\| \le M, \tag{2.24}$$

and satisfies the following gap conditions for a $\gamma \in (0,1)$

$$\begin{aligned} &\sup_{n\in\mathbb{Z}} \|(B_n^u)^{-1}\| < \gamma, \quad \sup_{n\in\mathbb{Z}} \|B_n\| \|(B_n^u)^{-1}\|^{k-\ell} \|B_n^{-1}\|^{\ell} < \gamma, \\ &\sup_{n\in\mathbb{Z}} \|B_n^s\| < \gamma, \quad \sup_{n\in\mathbb{Z}} \|B_n^{-1}\| \|B_n^s\|^{k-\ell} \|B_n\|^{\ell} < \gamma. \end{aligned} \tag{2.25}$$

(B2) Nonlinearity: $g_n \in C^k_{\mathrm{flat}}(\mathbb{R}^d)$ and additionally satisfies

$$\pi_u g_n(0, x^s) = 0 \quad \text{and} \quad \pi_s g_n(x^u, 0) = 0 \qquad (n \in \mathbb{Z}).$$

For $k, \ell \in \mathbb{N}$ and $B\colon \mathbb{Z} \times \mathbb{R}^d \to \mathbb{R}^d$, $(n,x) \mapsto B_n x$, with $B_n = \operatorname{diag}(B_n^u, B_n^s) \in \mathrm{Gl}(d,\mathbb{R})$, $B^u : \mathbb{Z} \to \mathbb{R}^{d_u\times d_u}$, $B^s : \mathbb{Z} \to \mathbb{R}^{d_s\times d_s}$, which satisfies (B1), define

$$\mathcal{D}^{k,\ell}_{\mathrm{flat}}(B) := \{x_{n+1} = B_n x_n + g_n(x_n) \colon g \text{ satisfies (B2)}\}.$$

Proposition 2.2 *If a system*

$$\dot{x} = A(t)x + f(t,x) \tag{2.26}$$

in $\mathcal{O}^k_{\text{flat}}(A)$, $k \geq 2$, *satisfies the spectral gap condition* (2.9) *for an* $\ell \in \mathbb{N}$, $\ell \leq k$, *i.e.*,

$$(k-\ell)\Sigma_u > \Sigma_u - \ell\Sigma_s \qquad \text{and} \qquad (k-\ell)\Sigma_s < \Sigma_s - \ell\Sigma_u, \tag{2.27}$$

then there exists a $\kappa > 0$ *such that for the associated time-*κ *system*

$$x_{n+1} = A_n^{(\kappa)} x_n + f_n^{(\kappa)}(x_n) \tag{2.28}$$

of (2.26), $A^{(\kappa)}$ *satisfies (B1) and* (2.28) *is in* $\mathcal{D}^{k,\ell}_{\text{flat}}(A^{(\kappa)})$.

Proof By Lemma 2.1, all time-κ systems

$$x_{n+1} = A_n^{(\kappa)} x_n + f_n^{(\kappa)}(x_n) \qquad (\kappa > 0)$$

associated with (2.26) satisfy (B2) and condition (2.24) of (B1). To conclude the proof, it is sufficient to find $\kappa > 0$ for which the linear part

$$A_n^{(\kappa)} = \Phi_A((n+1)\kappa, n\kappa) \tag{2.29}$$

satisfies condition (2.25) of (B1), where Φ_A denotes the evolution operator of the linear differential equation $\dot{\xi} = A(t)\xi = \text{diag}(A^u(t), A^s(t))\xi$. Then, $\Phi_A(\cdot,\cdot) = \text{diag}(\Phi_{A^u}(\cdot,\cdot), \Phi_{A^s}(\cdot,\cdot))$ and since $\Sigma_{\text{SS}}(A^u) = \Sigma_u > 0$ and $\Sigma_{\text{SS}}(A^s) = \Sigma_s < 0$, for any $\varepsilon > 0$, there exists $K > 0$ such that for all $t \geq s$

$$\begin{aligned} \|\Phi_A(t,s)\| &\leq K e^{(\max \Sigma_u + \varepsilon)(t-s)}, \\ \|\Phi_A(s,t)\| &\leq K e^{(-\min \Sigma_s + \varepsilon)(t-s)}, \\ \|\Phi_{A^u}(s,t)\| &\leq K e^{(-\min \Sigma_u + \varepsilon)(t-s)}. \end{aligned} \tag{2.30}$$

By assumption (2.27), there exists $\varepsilon > 0$ such that

$$\begin{aligned} (k-\ell)(\Sigma_u - \varepsilon) &> \Sigma_u + \varepsilon - \ell(\Sigma_s - \varepsilon), \\ (k-\ell)(\Sigma_s + \varepsilon) &< \Sigma_s - \varepsilon - \ell(\Sigma_u + \varepsilon). \end{aligned}$$

This, together with the estimates (2.30), implies for $t \geq s$

$$\|\Phi_A(t,s)\| \|\Phi_{A^u}(s,t)\|^{k-\ell} \|\Phi_A(s,t)\|^{\ell} \leq K^3 e^{-\beta(t-s)},$$

where

$$\beta := (\max \Sigma_u + \varepsilon) + (k-\ell)(-\min \Sigma_u + \varepsilon) + \ell(-\min \Sigma_s + \varepsilon) < 0.$$

Therefore, for $\gamma := \frac{1}{2}$, there exists $\kappa_1 > 0$ such that

$$\|\Phi_A(t,s)\|\,\|\Phi_{A^u}(s,t)\|^{k-\ell}\|\Phi_A(s,t)\|^{\ell} \le \gamma \quad \text{for } t-s \ge \kappa_1.$$

Similarly, there exists $\kappa_2 > 0$ such that

$$\|\Phi_A(s,t)\|\,\|\Phi_{A^s}(t,s)\|^{k-\ell}\|\Phi_A(t,s)\|^{\ell} \le \gamma \qquad \text{for } t-s \ge \kappa_2.$$

With $\kappa := \max\{\kappa_1, \kappa_2\} > 0$, the linear part (2.29) of the associated time-κ system of (2.26) satisfies condition (2.25) of (B1) and hence the associated time-κ system of (2.26) is in $\mathcal{D}^{k,\ell}_{\text{flat}}(A^{(\kappa)})$.

2.2.4 *Proof of Nonautonomous Sternberg Theorem*

In this subsection, we prove the Sternberg theorem 2.2 for nonautonomous differential equations. Recall that $A\colon \mathbb{R} \to \mathbb{R}^{d\times d}$ is of the form (2.10), i.e., $A(t) = \operatorname{diag}(A^u(t), A^s(t))$ with $0 \notin \Sigma_{\text{SS}}(A) = \Sigma^u \cup \Sigma^s$, $\Sigma^u = \Sigma_{\text{SS}}(A) \cap \mathbb{R}_{>0}$, $\Sigma^s = \Sigma_{\text{SS}}(A) \cap \mathbb{R}_{<0}$ and corresponding coordinates $x = (x^u, x^s) \in \mathbb{R}^{d_u} \times \mathbb{R}^{d_s}$. Define two closed subsets of $C^k_{\text{flat}}(\mathbb{R}^d)$:

$$C^k_{\text{flat, u}}(\mathbb{R}^d) := \{f \in C^k_{\text{flat}}(\mathbb{R}^d) \colon D^i f(0, x_s) = 0 \text{ for } i = 0, \dots, k\},$$

$$C^k_{\text{flat, s}}(\mathbb{R}^d) := \{f \in C^k_{\text{flat}}(\mathbb{R}^d) \colon D^i f(x_u, 0) = 0 \text{ for } i = 0, \dots, k\}.$$

Remark 2.2 In [32, Lemma 3], it is shown that for any $f \in C^k_{\text{flat}}(\mathbb{R}^d)$, there exists $\varphi \in C^k_{\text{flat, u}}(\mathbb{R}^d)$ and $\psi \in C^k_{\text{flat, s}}(\mathbb{R}^d)$ such that $f = \varphi + \psi$.

To prove the main Theorem 2.2, we show several preparatory results.

Proposition 2.3 (**C^k Equivalence of Discrete Time Systems with Close Nonlinearities**) *Let $\ell \in \mathbb{N}$, $1 \le \ell \le k$, and let B satisfy (B1) with $B_n = \operatorname{diag}(B^u_n, B^s_n) \in \operatorname{Gl}(d, \mathbb{R})$, $B^u \colon \mathbb{Z} \to \mathbb{R}^{d_u \times d_u}$, $B^s \colon \mathbb{Z} \to \mathbb{R}^{d_s \times d_s}$. Then two systems*

$$x_{n+1} = B_n x_n + f_n(x_n), \tag{2.31}$$

$$y_{n+1} = B_n y_n + g_n(y_n), \tag{2.32}$$

in $\mathcal{D}^{k,\ell}_{\text{flat}}(B)$ are C^k equivalent, if either

$$f_n - g_n \in C^k_{\text{flat,u}}(\mathbb{R}^d) \qquad (n \in \mathbb{N}), \tag{2.33}$$

or

$$f_n - g_n \in C^k_{\text{flat,s}}(\mathbb{R}^d) \qquad (n \in \mathbb{N}). \tag{2.34}$$

Note that we prove Proposition 2.3 only under the assumption that (2.33) holds. The proof under the assumption (2.34) is completely analogous. The main ingredient in the proof of Proposition 2.3 is the nonautonomous method of path which is developed in Sect. 2.2.5. To apply the method of path, we define the function $P\colon \mathbb{Z}\times\mathbb{R}^d\times[0,1]\to\mathbb{R}^d$, $(n,x,\tau)\mapsto P_n^\tau(x)$ by

$$P_n^\tau(x) := B_n x + (1-\tau)f_n(x) + \tau g_n(x) \tag{2.35}$$

and consider the homotopy of nonautonomous difference equations

$$z_{n+1} = P_n^\tau(z_n), \qquad \tau\in[0,1]$$

between (2.31) and (2.32). Actually, we do not apply the method of path directly to (2.31) and (2.32) but first replace the nonlinearities f_n and g_n by cut-off nonlinearities $\widetilde{f}_n := \psi\circ f_n$ and $\widetilde{g}_n := \psi\circ g_n$ where $\psi\colon\mathbb{R}^d\to\mathbb{R}^d$ is the C^∞ bump function

$$\psi(x) = \begin{cases} x, & \text{for } 0\le\|x\|\le r,\\ \exp\left(1-\frac{1}{1-(\|x\|-r)^2}\right)x, & \text{for } r<\|x\|<r+1,\\ 0, & \text{for } \|x\|\ge r, \end{cases} \tag{2.36}$$

for some $r>0$. This technical step does not change the differential equations (2.31) and (2.32) on a tubular neighborhood $\mathbb{R}\times B_r(0)$ of the zero solutions and ensures that P_n^τ is invertible. We use the notation

$$\begin{aligned}\operatorname{Diff}^k(\mathbb{R}^d) &:= \{f\in C^k(\mathbb{R}^d)\colon f^{-1}\text{ exists and } f^{-1}\in C^k(\mathbb{R}^d)\},\\ \operatorname{Diff}_0^k(\mathbb{R}^d) &:= \{f\in \operatorname{Diff}^k(\mathbb{R}^d)\colon f(0)=0\}.\end{aligned}$$

Lemma 2.2 (Invertibility of Homotopy Difference Equations) *For every $\delta>0$ there exists an $r>0$ such that, with* (2.31) *and* (2.32) *from Proposition 2.3 and ψ from* (2.36)*, the function $P\colon\mathbb{Z}\times\mathbb{R}^d\times[0,1]\to\mathbb{R}^d$, $(n,x,\tau)\mapsto P_n^\tau(x)$, defined by*

$$P_n^\tau(x) := B_n x + (1-\tau)\psi(f_n(x)) + \tau\psi(g_n(x)), \tag{2.37}$$

satisfies the following statements.

1. *$P_n^\tau\in\operatorname{Diff}_0^k(\mathbb{R}^d)$ for $n\in\mathbb{N}$ and the map $\mathbf{P}\colon[0,1]\to\operatorname{Diff}_0^k(\mathbb{R}^d)^{\mathbb{Z}}$, $\tau\mapsto\mathbf{P}^\tau := P_\cdot^\tau(\cdot)$ is continuous,*
2. *$\sup_{n\in\mathbb{Z},x\in\mathbb{R}^d}\|DP_n^\tau(x)-B_n\|\le\delta$, $\sup_{n\in\mathbb{Z},x\in\mathbb{R}^d}\|D(P_n^\tau)^{-1}(x)-B_n^{-1}\|\le\delta$.*

To prove Lemma 2.2, we need the following lemma.

Lemma 2.3 (Identity Plus C^k Contraction) *Let $\eta:\mathbb{R}^d\to\mathbb{R}^d$ be C^k with*

$$\rho := \sup_{x \in \mathbb{R}^d} \|D\eta(x)\| < 1. \tag{2.38}$$

Then, the map $f : \mathbb{R}^d \to \mathbb{R}^d$ defined by $f(x) := x + \eta(x)$ is a C^k diffeomorphism. Furthermore, for all $r > 0$, we have $B_{(1-\rho)r}(0) \subset f(B_r(0))$ and therefore

$$\|f^{-1}(x)\| \le \frac{1}{1-\rho}\|x\| \qquad \textit{for all } x \in \mathbb{R}^d.$$

Proof To show that f is injective, let $x, y \in \mathbb{R}^d$, $x \neq y$. By (2.38),

$$\|\eta(x) - \eta(y)\| < \|x - y\|,$$

which implies $\|f(x)-f(y)\| = \|x-y+\eta(x)-\eta(y)\| \ge \|x-y\|-\|\eta(x)-\eta(y)\| > 0$, i.e., $f(x) \neq f(y)$.

To show surjectivity of f, let $x \in \mathbb{R}^d$. Consider $T : \mathbb{R}^d \to \mathbb{R}^d$, $y \mapsto T(y)$, defined by

$$T(y) := x - \eta(y).$$

By (2.38),

$$\|T(y) - T(\widehat{y})\| = \|\eta(y) - \eta(\widehat{y})\| \le \rho\|y - \widehat{y}\|.$$

Then, by the Banach Contraction Mapping Theorem, there exists $y \in \mathbb{R}^d$ such that $T(y) = y$, or equivalently, $f(y) = x$. So far, we have shown that f is bijective.

Furthermore, by smoothness of η, the function f is also C^k. On the other hand, using the Inverse Function Theorem, the inverse f^{-1} of f is also C^k.

To conclude the proof, let $r > 0$ and $y \in B_{(1-\rho)r}(0)$. Let $x \in \mathbb{R}^d$ satisfy that $f(x) = y$. Hence, $\|x + \eta(x)\| = \|y\| < (1-\rho)r$ and therefore

$$\|x\| < (1-\rho)r + \|\eta(x)\| \le (1-\rho)r + \rho\|x\|,$$

which implies $\|x\| < r$, proving that $B_{(1-\rho)r}(0) \subset f(B_r(0))$.

Proof of Lemma 2.2 By (B1), there exists $M > 1$ such that

$$\sup_{n\in\mathbb{Z}} \|B_n\| \le M \quad \text{and} \quad \sup_{n\in\mathbb{Z}} \|B_n^{-1}\| \le M. \tag{2.39}$$

Let $\delta > 0$. Then there exists $\delta^* \in (0, \min\{\frac{\delta}{M}, \frac{1}{M^2}\})$ which additionally satisfies $\frac{M^3\delta^*}{1-M^2\delta^*} \le \delta$. Since $f_n, g_n \in C^k_{\text{flat}}(\mathbb{R}^d)$ for $n \in \mathbb{N}$, there exists an $r > 0$ such that with the bump function ψ from (2.36)

$$\sup_{n\in\mathbb{Z},x\in\mathbb{R}^d} \|D\psi(f_n(x))\| \le \delta^* \quad \text{and} \quad \sup_{n\in\mathbb{Z},x\in\mathbb{R}^d} \|D\psi(g_n(x))\| \le \delta^*. \tag{2.40}$$

For $\tau \in [0, 1]$, define

$$\eta_n^\tau(x) := (1-\tau)B_n^{-1}\psi(f_n(x)) + \tau B_n^{-1}\psi(g_n(x)).$$

Then, from (2.40), we derive that $\|D\eta_n^\tau(x)\| \le M\delta^* < 1$. Therefore, by virtue of Lemma 2.3, the map $x \mapsto x + \eta_n^\tau(x)$ is a C^k diffeomorphism. By (2.37), we have

$$P_n^\tau(x) = B_n(x + \eta_n^\tau(x)),$$

which implies that $P_n^\tau \in \mathrm{Diff}^k(\mathbb{R}^d)$. The continuous dependence of P_n^τ on τ follows directly from the definition (2.37) of P_n^τ and therefore (i) is proved.
(ii) For $\tau \in [0, 1]$, define

$$\Delta_n^\tau(x) := DP_n^\tau(x) - B_n = D\eta_n^\tau(x).$$

As in (i), from (2.39) and (2.40), we derive that

$$\|DP_n^\tau(x) - B_n\| \le M\delta^* \le \delta.$$

Also, $\|\Delta_n^\tau(x)B_n^{-1}\| \le M^2\delta^* < 1$. Then

$$\begin{aligned} DP_n^\tau(x)^{-1} &= (B_n + \Delta_n^\tau(x))^{-1} = B_n^{-1}(\mathrm{id} + \Delta_n^\tau B_n^{-1})^{-1} \\ &= B_n^{-1}\sum_{k=0}^{\infty}(-\Delta_n^\tau(x)B_n^{-1})^k \\ &= B_n^{-1}\Big(\mathrm{id} + \sum_{k=1}^{\infty}(-1)^k(\Delta_n^\tau(x)B_n^{-1})^k\Big), \end{aligned}$$

which implies that

$$\|DP_n^\tau(x)^{-1} - B_n^{-1}\| \le M\frac{M^2\delta^*}{1 - M^2\delta^*} \le \delta,$$

proving (ii).

To show Proposition 2.3, choose $r > 0$ such that the statements of Lemma 2.2 hold for P_n^τ, and additionally, using (B1), also the following inequalities hold for $n \in \mathbb{Z}$:

$$\left\|\frac{\partial(\pi_u \circ (P_n^\tau)^{-1})}{\partial x_u}\right\|_\infty < 1 \quad \text{and} \quad \|D_x(P_n^\tau)^{-1}\|_\infty > 1 \tag{2.41}$$

and

$$\|D_x P_n^\tau\|_\infty \left\|\frac{\partial(\pi_u \circ (P_n^\tau)^{-1}}{\partial x_u}\right\|_\infty^{k-\ell} \|D_x(P_n^\tau)^{-1}\|_\infty^\ell < \gamma,$$

$$\left\|D_x(P_n^\tau)^{-1}\right\|_\infty \left\|\frac{\partial(\pi_s \circ P_n^\tau)}{\partial x_u}\right\|_\infty^{k-\ell} \|D_x P_n^\tau\|_\infty^\ell < \gamma,$$

for a $\gamma \in (0, 1)$.

According to Theorem 2.4, to prove Proposition 2.3, it is sufficient to show the existence of $Z_n^\tau \in C_1^k(\mathbb{R}^d)$ such that

$$Z_{n+1}^\tau(x) - \frac{\partial P_n^\tau}{\partial x}\big((P_n^\tau)^{-1}(x)\big)Z_n^\tau\big((P_n^\tau)^{-1}(x)\big) = \frac{\partial P_n^\tau}{\partial \tau}\big((P_n^\tau)^{-1}(x)\big) \tag{2.42}$$

for $n \in \mathbb{N}$. In the following lemma, some properties of the right-hand side of (2.42) are established. For this purpose, let $\ell \in \mathbb{N}$, $1 \le \ell \le k$, $R > 0$, and define

$$\mathcal{M}_R^{k,\ell} := \left\{W \in C([0, 1], C_{\text{flat,u}}^\ell(\mathbb{R}^d)) : \left\|\tfrac{\partial^\ell W}{\partial x^\ell}(\tau, x)\right\| \le R\|\pi_u x\|^{k-\ell}\right\}.$$

Lemma 2.4 *The right-hand side*

$$X_n(\tau, x) := \frac{\partial P_n^\tau}{\partial \tau}\big((P_n^\tau)^{-1}(x)\big) \qquad (\tau \in [0, 1], n \in \mathbb{N}) \tag{2.43}$$

of (2.42) *satisfies*

1. $X_n(\tau, \cdot) \in C_{\text{flat,u}}^k(\mathbb{R}^d)$ *for* $n \in \mathbb{Z}$,
2. *there exists* $R > 0$ *such that* $X_n \in \mathcal{M}_R^{k,\ell}$ *for* $n \in \mathbb{Z}$.

Proof From (2.37) and (2.43), we derive that

$$X_n(\tau, x) = g_n\big((P_n^\tau)^{-1}(x)\big) - f_n\big((P_n^\tau)^{-1}(x)\big). \tag{2.44}$$

Since $f_n - g_n \in C_{\text{flat}}^k(\mathbb{R}^d)$, it follows that $X_n(\tau, \cdot) \in C_{\text{flat}}^k(\mathbb{R}^d)$. Furthermore, using (B2) the subspace $\{(0, x^s) \in \mathbb{R}^{d_u} \times \mathbb{R}^{d_s}\}$ is invariant under f_n and g_n. This together with (2.35) and the fact that $B_n = \operatorname{diag}(B_n^u, B_n^s)$ implies that

$$\begin{aligned}\pi_u P_n^\tau(0, x^s) &= \pi_u B_n(0, x^s)^{\mathrm{T}} + (1 - \tau)\pi_u f_n(0, x^s) + \tau\pi_u g_n(0, x^s)\\ &= 0.\end{aligned}$$

Consequently, the subspace $\{(0, x^s) \in \mathbb{R}^{d_u} \times \mathbb{R}^{d_s}\}$ is invariant under P_n^τ, i.e. $\pi_u(P_n^\tau)^{-1}(0, x^s) = 0$. By (2.33), we have $f_n - g_n \in C_{\text{flat,u}}^k(\mathbb{R}^d)$. Therefore,

$$D^i f_n\big((P_n^\tau)^{-1}(0, x^s)\big) - D^i g_n\big((P_n^\tau)^{-1}(0, x^s)\big) = 0.$$

Thus, from (2.44), we derive that $D_x^i X_n(\tau, 0, x^s) = 0$ and thus $X_n(\tau, \cdot) \in C^k_{\text{flat,u}}(\mathbb{R}^d)$.

(ii) Let

$$R := \sup_{x \in \mathbb{R}^d} \left\| \frac{\partial^k X_n}{\partial x^k}(\tau, x) \right\|.$$

Hence, for each $\tau \in [0, 1]$ using the Taylor expansion of the function $\frac{\partial^\ell X_n}{\partial x^\ell}(\tau, x)$ up to order $k - \ell$ at $(0, x^s)$ leads to

$$\left\| \frac{\partial^\ell X_n}{\partial x^\ell}(\tau, x) \right\| = \left\| \frac{\partial^\ell X_n}{\partial x^\ell}(\tau, x^u, x^s) - \frac{\partial^\ell X_n}{\partial x^\ell}(\tau, 0, x^s) \right\| \leq R \|\pi_u x\|^{k-\ell},$$

which completes the proof.

Next, we define for $n \in \mathbb{Z}$ the map $L_n \colon C\big([0, 1], C^\ell(\mathbb{R}^d)\big) \to C\big([0, 1], C^\ell(\mathbb{R}^d)\big)$ by

$$(L_n W)(\tau, x) := D_x P_n^\tau\big((P_n^\tau)^{-1}(x)\big)\, W\big(\tau, (P_n^\tau)^{-1}(x)\big). \tag{2.45}$$

Obviously, $\mathcal{M}_R^{k,\ell} \subset C([0, 1], C^\ell(\mathbb{R}^d))$ and in the following lemma, we provide some useful estimates on the restriction of the product of linear operators $L_n^j := L_{n+j-1} \cdots L_n$ on $\mathcal{M}_R^{k,\ell}$, where $n \in \mathbb{Z}$ and $j \in \mathbb{N}$.

Lemma 2.5 *For each $W \in \mathcal{M}_R^{k,\ell}$, the following statements hold for all $(\tau, x) \in [0, 1] \times \mathbb{R}^d$ and $n \in \mathbb{Z}$, $j \in \mathbb{N}$.*

1. $\|L_n^j W(\tau, x)\| \leq R\gamma^j \|\pi_u x\|^k$.
2. *There exists $\varepsilon > 0$ such that for $\|\pi_u x\| \leq \varepsilon$*

$$\left\| \frac{\partial^i L_n^j W}{\partial x^i}(\tau, x) \right\| \leq R\gamma^j \|\pi_u x\|^{k-i} \quad \textit{for } i = 1, \ldots, \ell.$$

Proof

(i) By (2.45), we have

$$L_n^j W(\tau, x) = \prod_{i=0}^{j-1} D_x P_{n+i}^\tau((P_{n+i}^\tau)^{-1}(x))\big(W(\tau, (P_{n+j-1}^\tau)^{-1} \circ \cdots \circ (P_n^\tau)^{-1}(x))\big),$$

which implies that

$$\|L_n^j W(\tau, x)\| \le \prod_{i=0}^{j-1} \|D_x P_{n+i}^\tau\|_\infty \|W(\tau, (P_{n+j-1}^\tau)^{-1} \circ \cdots \circ (P_n^\tau)^{-1}(x))\|. \tag{2.46}$$

Since $W \in \mathcal{M}_R^{k,\ell}$, it follows with the mean value theorem that

$$\|W(\tau, x)\| \le \frac{R}{(k-r+1)\cdots k} \|\pi_u x\|^k.$$

On the other hand, using the invariance of $\{0\} \times \mathbb{R}^{d_s}$ under P_n^τ and the mean value theorem, we obtain that

$$\|\pi_u \circ (P_{n+j-1}^\tau)^{-1} \circ \cdots \circ (P_n^\tau)^{-1}(x)\| \le \left(\prod_{i=0}^{j-1} \left\| \frac{\partial(\pi_u \circ (P_{n+i}^\tau)^{-1})}{\partial x_u} \right\|_\infty \right) \|\pi_u x\|.$$

Therefore, from (2.46), we derive that

$$\|L_n^j W(\tau, x)\| \le \frac{R}{(k-r+1)\cdots k} \|\pi_u x\|^k \prod_{i=0}^{j-1} \|D_x P_{n+i}^\tau\|_\infty \left\| \frac{\partial(\pi_u \circ (P_{n+i}^\tau)^{-1})}{\partial x_u} \right\|_\infty^k,$$

which together with (2.41) and (2.42) completes the proof of this part.

(ii) By induction, it is sufficient to show this assertion for $j = 1$, i.e., there exists $\varepsilon > 0$ such that for $\|\pi_u x\| \le \varepsilon$

$$\left\| \frac{\partial^i L_n W}{\partial x^i}(\tau, x) \right\| \le R\gamma \|\pi_u x\|^{k-i} \quad \text{for } i = 1, \dots, \ell.$$

Note that this assertion follows directly from [29, Lemma 4].

Proof of Proposition 2.3 By virtue of Lemma 2.5(i), the series

$$\sum_{j=0}^{\infty} L_{n-j}^j X_{n-1-j}(\tau, x) =: Z_n^\tau(x),$$

with $X_n(\tau, x)$ as in (2.43), converges for each $n \in \mathbb{Z}$. By definition of Z_n, we have

$$Z_{n+1}^\tau(x) = \sum_{j=0}^{\infty} L_{n+1-j}^j X_{n-j}(\tau, x)$$

$$= \sum_{j=1}^{\infty} L^j_{n+1-j} X_{n-j}(\tau, x) + X_n(\tau, x)$$

$$= L_n Z^\tau_n(x) + X_n(\tau, x),$$

which shows that $(Z^\tau_n(x))_{n\in\mathbb{Z}}$ is a solution of (2.42). Furthermore, using Lemma 2.5(ii), we get

$$\left\| \frac{\partial^i L^j_n X}{\partial x^i}(\tau, x) \right\| \leq L\gamma^j \|\pi_u x\|^{k-i} \quad \text{for } i = 1, \ldots, \ell \tag{2.47}$$

if $\|x\| \leq \varepsilon$ for some $\varepsilon > 0$, and by cutting off f_n and g_n, this estimate also holds for all $x \in \mathbb{R}^d$. Consequently, $Z^\tau_n(\cdot)$ is a C^ℓ function and in particular, $Z^\tau_n(0) = 0$ and $DZ^\tau_n(0) = 0$. Hence, $Z^\tau_n \in C^\ell_1(\mathbb{R}^d)$. Finally, by (2.47), the series $\sum_{j=0}^{\infty} L^j_{n+1-j} X_{n-j}(\tau, x)$ converges uniformly in τ to Z^τ_n in $\|\cdot\|_\ell$. Therefore, the map $\mathbf{Z}\colon [0, 1] \to C^\ell_1(\mathbb{R}^d)^{\mathbb{Z}}$, $\tau \mapsto \mathbf{Z}^\tau := Z^\tau_\cdot(\cdot)$, is continuous. By virtue of Theorem 2.4, the systems (2.31) and (2.32) are C^ℓ-equivalent.

We are now in a position to prove the main theorem.

Proof of Theorem 2.2 By Proposition 2.1, an arbitrary system in $\mathcal{O}^{k+2}(A)$ is C^k equivalent to a system

$$\dot{x} = A(t)x + g(x) \tag{2.48}$$

in $\mathcal{O}^k_{\text{flat}}(A)$. According to Proposition 2.2, there exists a $\kappa > 0$ such that the associated time-κ system of (2.48)

$$x_{n+1} = A^{(\kappa)}_n x_n + f^{(\kappa)}_n(x_n) \tag{2.49}$$

is an element of $\mathcal{D}^{k,\ell}_{\text{flat}}(A)$. By Remark 2.2, there exist $\varphi_n \in C^k_{\text{flat,u}}(\mathbb{R}^d)$ and $\psi_n \in C^k_{\text{flat,s}}(\mathbb{R}^d)$ for $n \in \mathbb{N}$ such that (2.49) can be rewritten as

$$x_{n+1} = A^{(\kappa)}_n x_n + \varphi_n(x_n) + \psi_n(x_n). \tag{2.50}$$

Since $\varphi_n \in C^k_{\text{flat,u}}(\mathbb{R}^d)$ for $n \in \mathbb{N}$, it follows together with Proposition 2.3 that system (2.50) is C^ℓ equivalent to the following system

$$x_{n+1} = A^{(\kappa)}_n x_n + \psi_n(x_n).$$

Applying Proposition 2.3 once again, this system is C^ℓ equivalent to the linear system

$$x_{n+1} = A^{(\kappa)}_n x_n.$$

Therefore, by Proposition 2.3, system (2.48) is C^ℓ equivalent to its linearization $\dot{x} = A(t)x$.

2.2.5 Method of Path

In this subsection, the *method of path* [61] is extended from autonomous to nonautonomous discrete time systems. It provides a way to construct a C^k equivalence between two nonautonomous difference equations

$$x_{n+1} = F_n(x_n), \tag{2.51}$$

$$y_{n+1} = G_n(y_n), \tag{2.52}$$

with $F_n, G_n \in \mathrm{Diff}^k(\mathbb{R}^d)$ for $n \in \mathbb{N}$, by considering the homotopy

$$P\colon \mathbb{Z} \times \mathbb{R}^d \times [0,1] \to \mathbb{R}^d, \quad (n, x, \tau) \mapsto P_n^\tau(x) := (1-\tau)F_n(x) + \tau G_n(x)$$

between F_n and G_n or, equivalently, the homotopy of difference equations

$$z_{n+1} = P_n^\tau(z_n), \qquad (\tau \in [0,1]) \tag{2.53}$$

between (2.51) and (2.52). Define $\mathbf{P}\colon [0,1] \to \mathrm{Diff}^k(\mathbb{R}^d)^{\mathbb{Z}}$, $\tau \mapsto \mathbf{P}^\tau := P_\cdot^\tau(\cdot)$.

Theorem 2.4 (Method of the Path) *Assume that*

1. $\mathbf{P}^\tau \in \mathrm{Diff}_0^k(\mathbb{R}^d)^{\mathbb{Z}}$ *for* $\tau \in [0,1]$,
2. $\mathbf{P}\colon [0,1] \to \mathrm{Diff}_0^k(\mathbb{R}^d)^{\mathbb{Z}}$ *is continuous,*
3. *For every* $\tau \in [0,1]$ *and* $n \in \mathbb{N}$, *there exists* $Z_n^\tau \in C_1^k(\mathbb{R}^d)$ *with*

$$Z_{n+1}^\tau(x) - \frac{\partial P_n^\tau}{\partial x}\big((P_n^\tau)^{-1}(x)\big) Z_n^\tau\big((P_n^\tau)^{-1}(x)\big) = \frac{\partial P_n^\tau}{\partial \tau}\big((P_n^\tau)^{-1}(x)\big) \tag{2.54}$$

for $n \in \mathbb{N}$ *and* $\mathbf{Z}\colon [0,1] \to C_1^k(\mathbb{R}^d)^{\mathbb{Z}}$, $\tau \mapsto \mathbf{Z}^\tau := Z_\cdot^\tau(\cdot)$, *is continuous.*

Then (2.51) *and* (2.52) *are* C^k *equivalent.*

Proof Since $\mathbf{Z} \in C([0,1], C_1^k(\mathbb{R}^d)^{\mathbb{Z}})$, there exists $K > 0$ such that

$$\|Z_n^\tau(x)\| \le K\|x\| \qquad \text{for all } n \in \mathbb{Z}, \tau \in [0,1], x \in \mathbb{R}^d,$$

Consequently, for $n \in \mathbb{Z}$, $x \in \mathbb{R}^d$ the initial value problem

$$\frac{d\xi}{d\tau}(\tau) = Z_n^\tau(\xi(\tau)), \qquad \xi(0) = x,$$

is uniquely solvable on [0, 1]. Let $T_n(\tau, x)$ denote its unique solution, i.e.,

$$\frac{\partial T_n}{\partial \tau}(\tau, x) = Z_n^\tau(T_n(\tau, x)), \qquad T_n(0, x) = x. \tag{2.55}$$

Then,

$$\|T_n(\tau, x)\| \leq \|x\| + M \int_0^\tau \|T_n(s, x)\| \, ds.$$

Using Gronwall's inequality [71, Chapter 3, Theorem 1.1], we obtain that $\|T_n(\tau, x)\| \leq e^{K\tau}\|x\|$ and thus $\lim_{x\to 0} T_n(\tau, x) = 0$ uniformly in $n \in \mathbb{Z}$.

To show that $(n, x) \mapsto T_n(\tau, x)$ is a C^k equivalence between (2.51) and (2.53), we first show that

$$T_{n+1}(\tau, F_n(x)) = P_n^\tau(T_n(\tau, x)) \qquad \text{for } n \in \mathbb{Z}, x \in \mathbb{R}^d. \tag{2.56}$$

For this purpose, let $n \in \mathbb{Z}$, $x \in \mathbb{R}^d$, and define $V : [0, 1] \times \mathbb{R}^d \to \mathbb{R}^d$ by

$$V(\tau, x) := P_n^\tau\big(T_n(\tau, F_n^{-1}(x))\big).$$

Taking derivatives on both sides with respect to τ yields

$$\frac{\partial V}{\partial \tau}(\tau, x) = D_x P_n^\tau\big(T_n(\tau, F_n^{-1}(x))\big)\frac{\partial T_n}{\partial \tau}(\tau, F_n^{-1}(x)) + \frac{\partial P_n^\tau}{\partial \tau}\big(T_n(\tau, F_n^{-1}(x))\big),$$

which, together with the fact that $T_n(\tau, F_n^{-1}(x)) = (P_n^\tau)^{-1}(V(\tau, x))$, implies

$$\frac{\partial V}{\partial \tau}(\tau, x) = D_x P_n^\tau\big((P_n^\tau)^{-1}(V(\tau, x))\big)\frac{\partial T_n}{\partial \tau}(\tau, F_n^{-1}(x)) + \frac{\partial P_n^\tau}{\partial \tau}\big((P_n^\tau)^{-1}(V(\tau, x))\big).$$

Using (2.55), we obtain

$$\begin{aligned}\frac{\partial V}{\partial \tau}(\tau, x) = {} & D_x P_n^\tau\big((P_n^\tau)^{-1}(V(\tau, x))Z_n(\tau, (P_n^\tau)^{-1}(V(\tau, x))) + \\ & + \frac{\partial P_n^\tau}{\partial \tau}((P_n^\tau)^{-1}(V(\tau, x)))\big),\end{aligned}$$

which, together with (2.54), gives

$$\frac{\partial V}{\partial \tau}(\tau, x) = Z_{n+1}(\tau, V(\tau, x)).$$

Furthermore, $V(0, x) = x$ and by the uniqueness of the solution of (2.55) for $n + 1$, we have $V(\tau, x) = T_{n+1}(\tau, x)$ and (2.56) follows.

To conclude the proof, it remains to show that $T_n(\tau, \cdot)$ is a local C^k diffeomorphism on $B_r(0)$ for some $r > 0$ which is independent of $n \in Z$. For this purpose, we define for $n \in \mathbb{N}$ the function $\eta_n(\tau, \cdot): \mathbb{R}^d \to \mathbb{R}^d$ by

$$\eta_n(\tau, x) := T_n(\tau, x) - x.$$

Since $T_n(0, x) = x$, it follows that $\eta_n(0, x) = 0$. Since $Z_n(\tau, x)$ is C^k in x, also $T_n(\tau, x)$ and $\eta_n(\tau, x)$ are C^k in x. Furthermore, from (2.55), we derive that

$$\frac{\partial D_x T_n}{\partial \tau}(\tau, x) = D_x Z_n(\tau, T_n(\tau, x)) D_x T_n(\tau, x),$$

which implies that

$$\frac{\partial D_x \eta_n}{\partial \tau}(\tau, x) = D_x Z_n(\tau, T_n(\tau, x))(\mathrm{id} + D_x \eta_n(\tau, x)),$$

or equivalently,

$$D_x \eta_n(\tau, x) = \int_0^\tau D_x Z_n(s, x + \eta_n(s, x))(\mathrm{id} + D_x \eta_n(s, x))\, ds. \tag{2.57}$$

Since $\mathbf{Z} \in C([0, 1], C_1^k(\mathbb{R}^d)^{\mathbb{Z}})$, there exists $L > 0$ such that $\|D_x Z_n(\tau, x)\| \le L\|x\|$. This, together with the fact that $\|T_n(\tau, x)\| \le e^{K\tau}\|x\| \le e^K\|x\|$, implies that

$$\|D_x Z_n(\tau, T_n(\tau, x))\| \le Le^K\|x\|.$$

Hence, from (2.57), we derive that

$$\begin{aligned}\|D_x \eta_n(\tau, x)\| &\le Le^K \int_0^\tau \|x\|(1 + \|D_x \eta_n(s, x)\|)\, ds \\ &\le Le^K\|x\| + Le^K\|x\| \int_0^\tau \|D_x \eta_n(s, x)\|\, ds.\end{aligned}$$

Using Gronwall's inequality [71, Chapter 3, Theorem 1.1], we obtain that

$$\|D_x \eta_n(\tau, x)\| \le Le^K\|x\| e^{Le^K\|x\|\tau}.$$

Consequently, there exists $r > 0$ such that

$$\rho := \sup_{n \in \mathbb{Z}} \max_{\tau \in [0,1]} \max_{\|x\| \le r} \|\eta_n(\tau, x)\| < 1.$$

In view of Lemma 2.3, the map $T_n(\tau, \cdot) : B_r(0) \to T_n(\tau, B_r(0))$ is a C^k diffeomorphism and $B_{(1-\rho)r}(0) \subset T_n(\tau, B_r(0))$. Furthermore, by Lemma 2.3, the inverse function of $T_n(\tau, \cdot)$ denoted by $S_n(\tau, \cdot) : B_{(1-\rho)r}(0) \to B_r(0)$ satisfies $\lim_{x\to 0} S_n(\tau, x) = 0$ uniformly in $n \in \mathbb{Z}$. Consequently, $(n, x) \mapsto T_n(\tau, x)$ is a C^k equivalence between (2.51) and (2.53) and, in particular, between (2.51) and (2.52) for $\tau = 1$.

2.3 Partial Linearization for Planar Nonautonomous Differential Equations

2.3.1 Partial Linearization Theorem

Consider a planar differential equation of the following form

$$\dot{x} = \begin{pmatrix} a_1(t) & 0 \\ 0 & a_2(t) \end{pmatrix} x + r(t, x) \tag{2.58}$$

with the zero solution, where $r : \mathbb{R} \times \mathbb{R}^2 \to \mathbb{R}^2$ is a smooth Carathéodory function, satisfying the following properties:

(P1) $r(t, 0) = 0$ and $D_x r(t, 0) = 0$.

(P2) The partial derivatives of r up to any finite order are globally bounded.

(P3) For $i = 1, 2$, the dichotomy spectrum of the equation $\dot{x}_i = a_i(t)x_i$ is $\lambda_i := [\underline{\lambda}_i, \overline{\lambda}_i]$ with $\lambda_2 < \lambda_1$.

Note that thanks to Theorem 2.1, any planar differential equations satisfying that the Sacker-Sell spectrum of the linear part consists of two intervals can be transformed to (2.58). Furthermore, let N be a nonnegative integer such that the following spectral gap condition holds

$$\lambda_2 \cap i\lambda_1 = \emptyset \qquad \text{for } i = 1, \ldots, N+1. \tag{2.59}$$

Then, by using nonautonomous invariant manifold theory (cf. Proposition 2.1), we can construct a finitely smooth equivalence to flatten the invariant manifold which corresponds to the spectral interval λ_2:

Theorem 2.5 (Flattening of an Invariant Manifold for Planar Nonautonomous Differential Equations) *Let N be an integer and assume that the dichotomy spectrum of* (2.58) *satisfies the following spectral gap condition* (2.59). *Then, system* (2.58) *is C^N equivalent to a system of the form*

$$\dot{y} = \begin{pmatrix} a_1(t) & 0 \\ 0 & a_2(t) \end{pmatrix} y + w(t, y)$$

where $w : \mathbb{R}\times\mathbb{R}^2 \to \mathbb{R}^2$ *is a* C^N *Carathéodory function which satisfies the following properties:*

(i) $w(t, 0) = 0$ *and* $D_y w(t, 0) = 0$*.*
(ii) All partial derivatives of w *in* y *up to order* N *are globally bounded.*
(iii) The set $\mathbb{R} \times \mathbb{R} \times \{0\}$ *is an invariant manifold, i.e., for* $t \in \mathbb{R}$ *and* $y_1 \in \mathbb{R}$ *we have* $w_2(t, y_1, 0) = 0$*.*

So far, we have mentioned that under a suitable spectral gap condition, a nonautonomous differential equation together with a reference solution can be transformed by a finitely smooth equivalence into a nonautonomous differential equation with the zero solution which satisfies that the linear part of the linearization is diagonal and the set $\mathbb{R} \times \mathbb{R} \times \{0\}$ is a flat invariant manifold. In this section, we take this situation as a starting point and consider the following system

$$\dot{x} = \begin{pmatrix} a_1(t) & 0 \\ 0 & a_2(t) \end{pmatrix} x + r(t, x) \tag{2.60}$$

where $r : \mathbb{R}\times\mathbb{R}^2 \to \mathbb{R}^2$ is a C^N Carathéodory function which satisfies the properties (i), (ii), and (iii) of Theorem 2.5. We now state the main result in the rest of this chapter.

Theorem 2.6 (Partial Linearization Theorem) *Consider system* (2.60)*, and let* $k \in \mathbb{N}$ *be arbitrary. Suppose that* $\lambda_2 < \lambda_1, \lambda_2 < 0$ *and that there exists* $n \in \mathbb{N}$ *such that* $4n - 3 \le N$ *and*

(A1) If $\overline{\lambda}_1 \ge 0$*, then* $\lambda_2 + (4n - 4)\lambda_1 < \lambda_1$ *and* $n\lambda_2 + k\lambda_1 - k\lambda_2 < 0$*.*
(A2) If $\overline{\lambda}_1 < 0$*, then* $n\lambda_2 - \lambda_2 < 0$ *and* $n\lambda_2 + \lambda_1 - k\lambda_2 < 0$*.*

Then system (2.60) *is* C^k *equivalent to the following partially linear system*

$$\dot{y} = \begin{pmatrix} a_1(t) & 0 \\ 0 & a_2(t) \end{pmatrix} y + \begin{pmatrix} p(t, y_1) \\ q(t, y_1)y_2 \end{pmatrix},$$

where the functions p *and* q *are defined as*

$$p(t, x_1) := r_1(t, x_1, 0), \quad q(t, x_1) := \frac{\partial r_2}{\partial x_2}(t, x_1, 0).$$

In the next subsections, we establish a proof of Theorem 2.6. The overview of the proof is:
Part I (Sect. 2.3.3): Our aim in this part is to establish some asymptotic estimate on the solution of the partially linearized system

$$\dot{x} = \begin{pmatrix} a_1(t) & 0 \\ 0 & a_2(t) \end{pmatrix} y + \begin{pmatrix} p(t, x_1) \\ q(t, x_1)x_2 \end{pmatrix}.$$

These estimates will be established in Sect. 2.3.3 and are used in the next two parts to construct suitable transformations.
Part II (Sect. 2.3.4): By definition of the functions p, q and property (P2), the function r can be written formally as

$$r(t, x) = \begin{pmatrix} p(t, x_1) \\ q(t, x_1)x_2 \end{pmatrix} + \begin{pmatrix} \sum_{i=1}^{n-1} \frac{1}{i!} \frac{\partial^i r_1}{\partial x_2^i}(t, x_1, 0)x_2^i \\ \sum_{i=2}^{n-1} \frac{1}{i!} \frac{\partial^i r_2}{\partial x_2^i}(t, x_1, 0)x_2^i \end{pmatrix} + O(|x_2|^n).$$

Our aim in this part is to eliminate step by step all terms

$$\sum_{i=1}^{n-1} \frac{1}{i!} \frac{\partial^i r_1}{\partial x_2^i}(t, x_1, 0)x_2^i \quad \text{and} \quad \sum_{i=2}^{n-1} \frac{1}{i!} \frac{\partial^i r_2}{\partial x_2^i}(t, x_1, 0)x_2^i$$

The required spectral gap condition for this task is $\lambda_2 + (4n-4)\overline{\lambda}_1 < \lambda_1$ in the case $\overline{\lambda}_1 \geq 0$. For the case $\overline{\lambda}_1 < 0$, we do not need any spectral gap condition for this procedure.

Note that in this part, we eliminate, informally speaking, an infinite number of nonresonant terms $x_1^j x_2^i$, $i = 1, \ldots, n-1$, $j = 1, \ldots, \infty$, at the same time. This is in contrast with the classical procedure in Poincaré-like results on hyperbolic systems, where nonresonant terms are removed order by order, i.e., by induction on the sum $i + j$. But here this is not sufficient: In Part III formulated below, we will need that low-order powers x_2^i are completely removed, i.e., for all j at once.
Part III (Sect. 2.3.5): As achieved in Part II, system (2.60) is C^n equivalent to the following system

$$\dot{x} = \begin{pmatrix} a_1(t) & 0 \\ 0 & a_2(t) \end{pmatrix} x + \begin{pmatrix} p(t, x_1) \\ q(t, x_1)x_2 \end{pmatrix} + O(|x_2|^n).$$

In this part, the remaining term $O(|x_2|^n)$ is eliminated under a C^k transformation. The spectral gap condition for this part is $n\lambda_2 + k\overline{\lambda}_1 - k\lambda_2 < 0$, resp. $n\lambda_2 - \lambda_2 < 0$ and $n\lambda_2 + \lambda_1 - k\lambda_2 < 0$, in the case $\overline{\lambda}_1 \geq 0$, resp. $\overline{\lambda}_1 < 0$.

2.3.2 *Estimate on the Derivatives for Solutions*

Our aim in this subsection is devoted to establishing an estimate on higher derivatives of solutions of the following perturbed system

$$\dot{x} = \begin{pmatrix} a_1(t) & 0 \\ 0 & a_2(t) \end{pmatrix} x + r(t, x), \tag{2.61}$$

where $r : \mathbb{R} \times \mathbb{R}^2 \to \mathbb{R}^2$ is a smooth Carathéodory function with globally bounded derivatives and satisfying that

$$r(t, 0) = 0 \quad \text{and} \quad \sup_{(t,x)\in\mathbb{R}\times\mathbb{R}^2} \|D_x r(t, x)\| \leq \frac{\delta}{2K(\varepsilon)},$$

where ε, δ are arbitrary positive numbers and $K(\varepsilon)$ is chosen such that inequality (2.6) holds. Before doing this, we need the following simple version of Bruno's formula. Note that Bruno's formula provides an explicit representation for higher-order derivatives of the composition of multivariable functions (see, e.g., [46, Theorem 2.1]). If the functions depend only on a scalar variable, then the formula becomes particularly simple (see, e.g., [70]).

Theorem 2.7 *Let $f, g : \mathbb{R} \to \mathbb{R}$ be smooth functions. Then,*

$$\frac{d^n}{dx^n} f \circ g(x) = \sum \frac{n!}{k_1! \cdots k_n!} f^{(k)}(g(x)) \left(\frac{g^{(1)}(x)}{1!}\right)^{k_1} \cdots \left(\frac{g^{(n)}(x)}{n!}\right)^{k_n},$$

where the sum is over all nonnegative integer solutions of the Diophantine equation $k_1 + 2k_2 + \cdots + nk_n = n$ and $k = k_1 + \cdots + k_n$.

Lemma 2.6 *Let $\varphi(t, s, x)$ denote the solution of (2.61) and let $n \in \mathbb{N}$. Then there exists a positive constant L such that for any $1 \leq i + j \leq n$ the following statements hold for all $x \in \mathbb{R}^2$:*

(i) If $\overline{\lambda}_1 \geq 0$, we have $\left\| \frac{\partial^{i+j}\varphi}{\partial x_1^i \partial x_2^j}(t, s, x) \right\| \leq Le^{(i+j)(\overline{\lambda}_1+\varepsilon+\delta)(t-s)}$ for $t \geq s$.

(ii) If $\overline{\lambda}_1 < 0$ and assume additionally that $\overline{\lambda}_1+\varepsilon+\delta < 0$, then $\left\| \frac{\partial^{i+j}\varphi}{\partial x_1^i \partial x_2^j}(t, s, x) \right\| \leq Le^{(\overline{\lambda}_1+\varepsilon+\delta)(t-s)}$ for $t \geq s$.

(iii) $\left\| \frac{\partial^{i+j}\varphi}{\partial x_1^i \partial x_2^j}(t, s, x) \right\| \leq Le^{(i+j)(\underline{\lambda}_2-\varepsilon-\delta)(t-s)}$ for $t \leq s$.

Proof (i) In fact, we prove by induction on $n = i + j$ the following stronger inequality for $x \in \mathbb{R}^2$

$$\left\| \frac{\partial^{i+j}\varphi}{\partial x_1^i \partial x_2^j}(t, s, x) \right\| \leq Le^{\left((i+j)(\overline{\lambda}_1+\varepsilon+\delta)-\frac{\delta}{2}\right)(t-s)} \quad \text{for } t \geq s. \tag{2.62}$$

By (2.61),

$$\varphi(t, s, x) = \Lambda(t, s)x + \int_s^t \Lambda(t, u)r(u, \varphi(u, s, x))\, du, \tag{2.63}$$

which implies that

$$D_x\varphi(t, s, x) = \Lambda(t, s) + \int_s^t \Lambda(t, u)D_x r(u, \varphi(u, s, x))D_x\varphi(u, s, x)\, du.$$

Therefore,

$$\|D_x\varphi(t, s, x)\| \le K(\varepsilon)e^{(\overline{\lambda}_1+\varepsilon)(t-s)} + \frac{\delta}{2}\int_s^t e^{(\overline{\lambda}_1+\varepsilon)(t-u)}\|D_x\varphi(u, s, x)\|\, du,$$

which implies that

$$\|e^{(\overline{\lambda}_1+\varepsilon)(s-t)}D_x\varphi(t, s, x)\| \le K(\varepsilon) + \frac{\delta}{2}\int_s^t \|e^{(\overline{\lambda}_1+\varepsilon)(s-u)}D_x\varphi(u, s, x)\|\, du.$$

Applying Gronwall's inequality yields

$$\|e^{(\overline{\lambda}_1+\varepsilon)(s-t)}D_x\varphi(t, s, x)\| \le K(\varepsilon)e^{\frac{\delta}{2}(t-s)} \quad \text{for } t \ge s,$$

which proves (2.62) for $n = 1$. Suppose that the assertion holds for n, i.e., there exists L such that for all $i, j = 0, \ldots, n$ with $1 \le i + j \le n$, we have

$$\left\| \frac{\partial^{i+j}\varphi}{\partial x_1^i \partial x_2^j}(t, s, x) \right\| \le Le^{\left((i+j)(\overline{\lambda}_1+\varepsilon+\delta)-\frac{\delta}{2}\right)(t-s)}. \tag{2.64}$$

Choose and fix $i, j \in \{0, \ldots, n+1\}$ with $i + j = n + 1$. By (2.64) and Bruno's formula in Theorem 2.7, we get for $u \ge s$

$$\frac{d^{n+1}}{dx_1^i dx_2^j}(r(u, \varphi(u, s, x))) = D_x r(u, \varphi(u, s, x))\frac{\partial^{n+1}\varphi}{\partial x_1^i \partial x_2^j}(u, s, x) + R(u, s, x),$$

where the term $R(u, s, x)$ can be estimated as

$$\|R(u, s, x)\| \le Ce^{((n+1)(\overline{\lambda}_1+\varepsilon+\delta)-\delta)(u-s)}.$$

Using (2.63), we obtain

$$\frac{\partial^{n+1}\varphi}{\partial x_1^i \partial x_2^j}(t, s, x) = \int_s^t \Lambda(t, u)\left(D_x r(u, \varphi(u, s, x))\frac{\partial^{n+1}}{\partial x_1^i \partial x_2^j}(u, s, x) \right.$$
$$\left. + R(u, s, x)\right) du.$$

Consequently,

$$\left\|\frac{\partial^{n+1}\varphi}{\partial x_1^i \partial x_2^j}(t,s,x)\right\| \le \frac{\delta}{2}\int_s^t e^{(\overline{\lambda}_1+\varepsilon)(t-u)}\left\|\frac{\partial^{n+1}\varphi}{\partial x_1^i \partial x_2^j}(u,s,x)\right\| du + CK(\varepsilon)\int_s^t e^{(\overline{\lambda}_1+\varepsilon)(t-u)}e^{((n+1)(\overline{\lambda}_1+\varepsilon+\delta)-\delta)(u-s)}\, du.$$

Hence,

$$\left\|e^{-(\overline{\lambda}_1+\varepsilon)(t-s)}\frac{\partial^{n+1}\varphi}{\partial x_1^i \partial x_2^j}(t,s,x)\right\| \le \frac{\delta}{2}\int_s^t \left\|e^{-(\overline{\lambda}_1+\varepsilon)(u-s)}\frac{\partial^{n+1}\varphi}{\partial x_1^i \partial x_2^j}(u,s,x)\right\| du + \frac{CK(\varepsilon)}{n(\overline{\lambda}_1+\varepsilon+\delta)}e^{n(\overline{\lambda}_1+\varepsilon+\delta)(t-s)},$$

which implies that

$$\left\|e^{-(\overline{\lambda}_1+\varepsilon)(t-s)}\frac{\partial^{n+1}\varphi}{\partial x_1^i \partial x_2^j}(t,s,x)\right\| \le \frac{CK(\varepsilon)}{n(\overline{\lambda}_1+\varepsilon+\delta)}e^{\left(\frac{\delta}{2}+n(\overline{\lambda}_1+\varepsilon+\delta)\right)(t-s)}.$$

Multiplying both sides with $e^{(\overline{\lambda}_1+\varepsilon)(t-s)}$ yields (2.62) for $i+j=n+1$, and the proof of this part is complete.

2.3.3 *Estimate on the Derivatives for Solutions of Partially Linear Systems*

Let $p, q : \mathbb{R}\times\mathbb{R} \to \mathbb{R}$ be smooth Carathéodory functions such that all partial derivatives of p and q in x are globally bounded and for $t \in \mathbb{R}$

$$|p(t,x_1)|, |q(t,x_1)| \le \delta \quad \text{and} \quad \left|\frac{\partial p}{\partial x_1}(t,x_1)\right| \le \frac{\delta}{2K(\varepsilon)}, \tag{2.65}$$

where ε, δ are arbitrary positive numbers and $K(\varepsilon)$ is chosen such that inequality (2.6) holds. Not that this estimate can be done by using the cutoff technique (Example 2.3). Additionally, in the case $\overline{\lambda}_1 < 0$, the positive constants ε, δ are assumed to satisfy $\overline{\lambda}_1+\varepsilon+\delta < 0$.

Consider the following partially linear system

$$\begin{aligned}\dot{x}_1 &= a_1(t)x_1 + p(t,x_1),\\ \dot{x}_2 &= a_2(t)x_2 + q(t,x_1)x_2.\end{aligned} \tag{2.66}$$

Let $\Phi(\cdot, s, x_1, x_2)$ denote the unique solution of (2.66) which satisfies that $x(s) = (x_1, x_2)^{\mathrm{T}}$. In the following lemma, we state some properties of $\Phi =: (\Phi_1, \Phi_2)$.

Lemma 2.7 *The following statements hold:*

(i) $\Phi_1(t, s, x_1, x_2)$ *is independent of* x_2 *and to shorten the notation, we abbreviate it as* $\Phi_1(t, s, x_1)$.

(ii) $\Phi_2(t, s, x_1, x_2) = \Lambda_2(t, s) x_2 \exp\left(\int_s^t q(u, \Phi_1(u, s, x_1))\, du\right)$.

Proof The proof of (i) is trivial. To show statement (ii), we define a new variable $z := \ln x_2$. From (2.66), we have

$$\dot{z} = a_2(t) + q(t, x_1),$$

which implies that

$$z(t) = z(s) + \int_s^t a_2(u) + q(u, x_1(u))\, du.$$

Replacing $x_2 = e^z$ and using the fact that $\Lambda_2(t, s) = \exp\left(\int_s^t a_2(u)\, du\right)$, we obtain

$$x_2(t) = \Lambda_2(t, s) x_2 \exp\left(\int_s^t q(u, x_1(u))\, du\right),$$

which completes the proof.

An estimate on higher-order partial derivatives of Φ_1 in forward time is obtained as follows.

Lemma 2.8 *There exists a positive constant* L *such that for any* $1 \le i + j \le N$, *the following statements hold for all* $t \ge s$ *and* $x \in \mathbb{R}^2$:

(i) If $\overline{\lambda}_1 \ge 0$, *we have* $\left\| \frac{\partial^{i+j} \Phi_1}{\partial x_1^i \partial x_2^j}(t, s, x) \right\| \le L e^{(i+j)(\overline{\lambda}_1 + \varepsilon + \delta)(t-s)}$.

(ii) If $\overline{\lambda}_1 < 0$, *we have* $\left\| \frac{\partial^{i+j} \Phi_1}{\partial x_1^i \partial x_2^j}(t, s, x) \right\| \le L e^{(\overline{\lambda}_1 + \varepsilon + \delta)(t-s)}$.

Proof Using the fact that $\left\| \frac{\partial^{i+j} \Phi_1}{\partial x_1^i \partial x_2^j}(t, s, x) \right\| \le \left\| \frac{\partial^{i+j} \Phi}{\partial x_1^i \partial x_2^j}(t, s, x) \right\|$, applying Lemma 2.6 with $r(t, x) := \begin{pmatrix} p(t, x_1) \\ 0 \end{pmatrix}$ yields the claim.

In light of Lemma 2.7(ii), $\Phi_2(t, s, x)$ can be represented as a function of $\Phi_1(t, s, x_1)$. Combining this fact and the preceding Lemma yields the following estimate on the higher derivatives of Φ_2:

Lemma 2.9 *There exists a positive constant* L *such that the following statements hold for all* $t \ge s$ *and* $x \in \mathbb{R}^2$:

(i) If $\overline{\lambda}_1 > 0$, *we have*

$$\left|\frac{\partial^i \Phi_2}{\partial x_1^i}(t,s,x_1,x_2)\right| \le Le^{\left((\overline{\lambda}_2+\varepsilon+\delta)+i(\overline{\lambda}_1+\varepsilon+\delta)\right)(t-s)}|x_2| \quad \textit{for } i=0,\dots,N,$$

$$\left|\frac{\partial^{i+1} \Phi_2}{\partial x_1^i \partial x_2}(t,s,x_1,x_2)\right| \le Le^{\left((\overline{\lambda}_2+\varepsilon+\delta)+i(\overline{\lambda}_1+\varepsilon+\delta)\right)(t-s)} \quad \textit{for } i=0,\dots,N-1.$$

(ii) If $\overline{\lambda}_1 < 0$, *we have*

$$\left|\frac{\partial^i \Phi_2}{\partial x_1^i}(t,s,x_1,x_2)\right| \le Le^{\left((\overline{\lambda}_2+\varepsilon+\delta)\right)(t-s)}|x_2| \quad \textit{for } i=0,\dots,N,$$

$$\left|\frac{\partial^{i+1} \Phi_2}{\partial x_1^i \partial x_2}(t,s,x_1,x_2)\right| \le Le^{\left((\overline{\lambda}_2+\varepsilon+\delta)\right)(t-s)} \quad \textit{for } i=0,\dots,N-1.$$

2.3.4 Normalization Up to Finite Order Along the Flat Invariant Manifold

In the first part of this subsection, we will define some constants which will be used throughout this subsection. Depending on the sign of $\overline{\lambda}_1$, we need a gap condition to be satisfied:

$$\overline{\lambda}_2 - \underline{\lambda}_1 + (4n-4)\overline{\lambda}_1 + (4n-2)(\varepsilon+\delta) < 0, \qquad \text{if } \overline{\lambda}_1 \ge 0, \tag{2.67}$$

and

$$\overline{\lambda}_2 - \underline{\lambda}_1 + \overline{\lambda}_1 + (4n-2)(\varepsilon+\delta) < 0, \qquad \text{if } \overline{\lambda}_1 < 0. \tag{2.68}$$

Condition (2.67) follows from the spectral gap condition $\lambda_2 + (4n-4)\lambda_1 < \lambda_1$ in Theorem 2.6 if $\overline{\lambda}_1 > 0$ and condition (2.68) is automatically satisfied if $\overline{\lambda}_1 < 0$. Choose and fix ε, δ satisfying the (2.67), resp. (2.68). Using cutoff functions and the fact that $p(t,0) = \frac{\partial p}{\partial x}(t,0) = q(t,0) = 0$, we can assume that p and q satisfy (2.65). Therefore, all inequalities in Lemmas 2.8 and 2.9 hold.

Before stating and proving the main result in this subsection, we introduce a class of Carathéodory functions which can be written as the sum of certain Carathéodory functions of higher smoothness order.

Definition 2.4 Let $m, n \in \mathbb{N}$ with $m \ge n$. A Carathéodory function $\alpha : \mathbb{R} \times \mathbb{R}^2 \to \mathbb{R}^2$ is said to be in class $C(m,n)$ if there exist C^{m-j} Carathéodory functions $\alpha_j(t,x_1)$, for $j = 0,\dots,n-1$, such that the function

$$\widehat{\alpha}(t,x) := \alpha(t,x) - \sum_{j=0}^{n-1} \alpha_j(t,x_1)x_2^j$$

is a C^{m-n} Carathéodory function and $\widehat{\alpha}(t,x) = O(x_2^n)$.

Of course any C^m Carathéodory function $\alpha(t,x)$ is of class $C(m,n)$ with $n \leq m$ and the corresponding α_j are just derived from Taylor's theorem. Inversely, A $C(m,n)$ function is of class C^{m-n}; in other words $C^m \subset C(m,n) \subset C^{m-n}$ for $n < m$. The class $C(m,n)$ is introduced to accurately trace smoothness information of a function after applying Taylor's theorem to it.

Proposition 2.4 *Let $\ell \in \{1,\dots,n-1\}$ and $m \geq n$ be arbitrary. Consider the following system*

$$\dot{x} = \begin{pmatrix} a_1(t) & 0 \\ 0 & a_2(t) \end{pmatrix} x + \begin{pmatrix} p(t,x_1) \\ q(t,x_1)x_2 \end{pmatrix} + \begin{pmatrix} b(t,x_1)x_2^\ell \\ 0 \end{pmatrix} + r(t,x), \tag{2.69}$$

where $b(t,x_1)$ is a $C^{m+1-\ell}$ Carathéodory function and $r : \mathbb{R} \times \mathbb{R}^2 \to \mathbb{R}^2$ is in class $C(m+1,n)$ and satisfies $r(t,x) = O(x_2^{\ell+1})$. Suppose the following spectral gap condition holds:

(i) If $\overline{\lambda}_1 \geq 0$, then $\ell\overline{\lambda}_2 - \underline{\lambda}_1 + (m+1-\ell)\overline{\lambda}_1 + (m+2\ell)(\varepsilon+\delta) < 0$.
(ii) If $\overline{\lambda}_1 < 0$, then $\ell\overline{\lambda}_2 - \underline{\lambda}_1 + \overline{\lambda}_1 + (m+2)(\varepsilon+\delta) < 0$.

Then, system (2.69) is $C^{m+1-\ell}$ equivalent to a system of the following form

$$\dot{y} = \begin{pmatrix} a_1(t) & 0 \\ 0 & a_2(t) \end{pmatrix} y + \begin{pmatrix} p(t,y_1) \\ q(t,y_1)y_2 \end{pmatrix} + w(t,y), \tag{2.70}$$

where $w(t,y)$ is in class $C(m,n)$ and satisfies $w(t,y) = O(|y_2|^{\ell+1})$.

Proof Let $\varphi(t,s,x_1,x_2)$ denote the solution of (2.69) with the initial condition $x(s) = (x_1,x_2)^{\mathrm{T}}$. Recall that $\Lambda(t,s)x$, resp. $\Phi(t,s,x)$, denote the solutions satisfying the initial value problem $x(s) = x$ of the following systems

$$\dot{x} = \begin{pmatrix} a_1(t) & 0 \\ 0 & a_2(t) \end{pmatrix} x, \quad \text{resp.} \quad \dot{x} = \begin{pmatrix} a_1(t) & 0 \\ 0 & a_2(t) \end{pmatrix} x + \begin{pmatrix} p(t,x_1) \\ q(t,x_1)x_2 \end{pmatrix}.$$

In order to construct a smooth transformation between system (2.69) and system (2.70), we define a scalar function $h_{s,t} : \mathbb{R} \to \mathbb{R}$ for $s,t \in \mathbb{R}$ by

$$h_{s,t}(x_1) := \lim_{x_2 \to 0} \frac{\varphi_1(s,t) \circ \Phi(t,s)(x_1,x_2) - x_1}{x_2^\ell} \qquad \text{for } x_1 \in \mathbb{R}.$$

In the following steps, we investigate some properties of the function $h_{s,t}$ and explain how to construct the desired transformation by using the function $h_{s,t}$.
Step 1: In this step, we prove that

$$h_{s,t}(x_1) = \int_s^t \Lambda_1(s,u)\Lambda_2(u,s)^\ell g_{u,s}(x_1)\,du, \tag{2.71}$$

where

$$g_{u,s}(x_1) := b(u, \Phi_1(u,s,x_1)) \frac{\exp\left(\ell \int_s^u q(v, \Phi_1(v,s,x_1))\,dv\right)}{\exp\left(\int_s^u \frac{\partial p}{\partial x_1}(v, \Phi_1(v,s,x_1))\,dv\right)}. \tag{2.72}$$

For this purpose, for each $\tau, t, s, x_1 \in \mathbb{R}$, we define

$$R(\tau,t,s,x_1) := \lim_{x_2 \to 0} \frac{\varphi_1(\tau,t) \circ \Phi(t,s)(x_1,x_2) - \Phi_1(\tau,s,x_1)}{x_2^\ell}.$$

Since φ is a solution of (2.69), it follows that

$$\begin{aligned}\frac{d}{d\tau}(\varphi_1(\tau,t) \circ \Phi(t,s)(x)) &= a_1(\tau)\varphi_1(\tau,t) \circ \Phi(t,s)(x) \\ &\quad + p(\tau, \varphi_1(\tau,t) \circ \Phi(t,s)(x)) \\ &\quad + b(\tau, \varphi_1(\tau,t) \circ \Phi(t,s)(x)) \\ &\quad \times (\varphi_2(\tau,t) \circ \Phi(t,s)(x))^\ell + O(|x_2|^{\ell+1}).\end{aligned} \tag{2.73}$$

By definition of $\Phi_1(\tau,s,x_1)$, we also have

$$\frac{d}{d\tau}\Phi_1(\tau,s,x_1) = a_1(\tau)\Phi_1(\tau,s,x_1) + p(\tau, \Phi_1(\tau,s,x_1)). \tag{2.74}$$

Using the mean value theorem, we obtain

$$\begin{aligned}&\lim_{x_2 \to 0} \frac{p(\tau, \varphi_1(\tau,t) \circ \Phi(t,s)(x)) - p(\tau, \Phi_1(\tau,s,x_1))}{x_2^\ell} \\ &= \frac{\partial p}{\partial x_1}(\tau, \Phi_1(\tau,s,x_1)) \lim_{x_2 \to 0} \frac{\varphi_1(\tau,t) \circ \Phi(t,s)(x) - \Phi_1(\tau,s,x_1)}{x_2^\ell} \\ &= \frac{\partial p}{\partial x_1}(\tau, \Phi_1(\tau,s,x_1)) R(\tau,t,s,x_1).\end{aligned}$$

Using $\lim_{x_2 \to 0} \varphi_1(\tau,t) \circ \Phi(t,s)(x) = \Phi_1(\tau,s,x_1)$ and subtracting each sides of (2.73), (2.74) and then dividing by x_2^ℓ yields that

$$\frac{d}{d\tau}R(\tau,t,s,x_1) = \Big(a_1(\tau) + \frac{\partial p}{\partial x_1}(\tau, \Phi_1(\tau,s,x_1))\Big) R(\tau,t,s,x_1) +$$
$$b(\tau, \Phi_1(\tau,s,x_1)) \lim_{x_2\to 0}\Big(\frac{\varphi_2(\tau,t)\circ\Phi(t,s)(x_1,x_2)}{x_2}\Big)^{\ell}.$$

Thus, to further simplify the expression of $\frac{d}{d\tau}R(\tau,t,s,x_1)$, we need to compute $\lim_{x_2\to 0}\frac{\varphi_2(\tau,t)\circ\Phi(t,s)(x_1,x_2)}{x_2}$. Using a similar technique as in the proof of Lemma 2.7(ii), we obtain that for $t, s \in \mathbb{R}$ and $(x_1,x_2)^{\mathrm{T}} \in \mathbb{R}^2$

$$\varphi_2(t,s)(x_1,x_2) = \Lambda_2(t,s)x_2 \exp\Big(\int_s^t q(u, \varphi_1(u,s,x_1,x_2))\, du\Big) + O(|x_2|^{\ell+1}),$$

which together with Lemma 2.7(ii) implies that

$$\begin{aligned}
&\varphi_2(\tau,t)\circ\Phi(t,s)(x_1,x_2)\\
&= \Lambda_2(\tau,t)\Phi_2(t,s,x_1,x_2)\exp\Big(\int_t^\tau q(u, \varphi_1(u,t)\circ\Phi(t,s,x_1,x_2))\, du\Big) + O(|x_2|^{\ell+1})\\
&= \Lambda_2(\tau,s)x_2\exp\Big(\int_s^t q(u, \Phi_1(u,s,x_1))\, du\Big)\\
&\quad \exp\Big(\int_t^\tau q(u, \varphi_1(u,t)\circ\Phi(t,s)(x_1,x_2))\, du\Big) + O(|x_2|^{\ell+1}).
\end{aligned}$$

Consequently,

$$\lim_{x_2\to 0}\frac{\varphi_2(\tau,t)\circ\Phi(t,s)(x_1,x_2)}{x_2} = \Lambda_2(\tau,s)\exp\Big(\int_s^\tau q(u, \Phi_1(u,s,x_1))\, du\Big).$$

Therefore,

$$\frac{d}{d\tau}R(\tau,t,s,x_1) = \Big(a_1(\tau) + \frac{\partial p}{\partial x_1}(\tau, \Phi_1(\tau,s,x_1))\Big) R(\tau,t,s,x_1) +$$
$$b(\tau, \Phi_1(\tau,s,x_1))\Lambda_2(\tau,s)^{\ell}\exp\Big(\ell\int_s^\tau q(u, \Phi_1(u,s,x_1))\, du\Big).$$

Consequently, by the variation of constants formula and the fact that $R(t,t,s,x_1) = 0$, we get (2.71).

Step 2: In this step, we show the existence of the limit function $h_s : \mathbb{R} \to \mathbb{R}$ defined by $h_s(x_1) = \lim_{t\to\infty} h_{s,t}(x_1)$. For this purpose, let $t_2 > t_1 > s$ be arbitrary. Using (2.71), we obtain that

$$|h_{s,t_2}(x_1) - h_{s,t_1}(x_1)| \le \int_{t_1}^{t_2} \Lambda_1(s,u)\Lambda_2(u,s)^{\ell}|g_{u,s}(x_1)|\, du,$$

which implies that

$$|h_{s,t_2}(x_1) - h_{s,t_1}(x_1)| \le K(\varepsilon)^{\ell+1} \int_{t_1}^{t_2} e^{(\ell(\overline{\lambda}_2+\varepsilon)-\underline{\lambda}_1+\varepsilon)(u-s)} |g_{u,s}(x_1)|\, du.$$

Note that by (2.65), the definition of g in (2.72) and the boundedness of the derivatives of p and q in x, we get that

$$|g_{u,s}(x_1)| \le Me^{(\ell+1)\delta(u-s)} \qquad \text{for } u \ge s.$$

Therefore,

$$|h_{s,t_2}(x_1) - h_{s,t_1}(x_1)| \le MK(\varepsilon)^{\ell+1} \int_{t_1}^{t_2} e^{(\ell\overline{\lambda}_2-\underline{\lambda}_1+(\ell+1)(\varepsilon+\delta))(u-s)}\, du,$$

which together with the fact that $\ell\overline{\lambda}_2 - \underline{\lambda}_1 + (\ell+1)(\varepsilon+\delta) < 0$ implies that the limit $\lim_{t\to\infty} h_{s,t}(x_1)$ exists uniformly in x_1.

Step 3: At this moment, there is enough information to define the required equivalence relation, but we prefer to discuss the smoothness in this and the next step. The equivalence relation will then be dealt with in Step 5. In this step, we show that h_s is $C^{m+1-\ell}$. It is sufficient to show the uniform convergence in x_1 of the following improper integral

$$\int_s^{\infty} \Lambda_1(s,u)\Lambda_2(u,s)^{\ell} \frac{d^i g_{u,s}}{dx_1^i}(x_1)\, du \qquad \text{for } i = 1, \dots, m+1-\ell. \tag{2.75}$$

Thereto, we consider the following two cases $\overline{\lambda}_1 \ge 0$ and $\overline{\lambda}_1 < 0$.

<u>Case 1</u>: $\overline{\lambda}_1 \ge 0$. Using Lemma 2.8 and Theorem 2.7, it is seen that there is a positive constant C_1 such that for $u \ge s$

$$\left|\frac{d^i g_{u,s}}{\partial x_1^i}(x_1)\right| \le C_1 e^{\left((\ell+1)\delta + i(\overline{\lambda}_1+\varepsilon+\delta)\right)(u-s)} \qquad \text{for } i = 1, \dots, m+1-\ell.$$

Consequently, for any $t_2 > t_1 > s$ and $i = 1, \dots, m+1-\ell$, we get

$$\left|\int_{t_1}^{t_2} \Lambda_1(s,u)\Lambda_2(u,s)^{\ell} \frac{d^i g_{u,s}}{\partial x_1^i}(x_1)\, du\right| \le \widehat{C}_1 \frac{e^{(t_2-s)\alpha_1} - e^{(t_1-s)\alpha_1}}{\alpha_1},$$

where

$$\widehat{C}_1 := C_1 K(\varepsilon)^{\ell+1}, \quad \alpha_1 := \ell\overline{\lambda}_2 - \underline{\lambda}_1 + i\overline{\lambda}_1 + (\ell+i+1)(\varepsilon+\delta).$$

Due to the fact that ε, δ satisfy the spectral gap condition (i), the uniform convergence in x_1 in (2.75) is shown. Therefore, the function h_s is $C^{m+1-\ell}$.
<u>Case 2</u>: $\overline{\lambda}_1 < 0$. Using Lemma 2.8 and Corollary 2.7, it is seen that there is a positive constant C_2 such that for $u \geq s$

$$\left|\frac{d^i g_{u,s}}{\partial x_1^i}(x_1)\right| \leq C_2 e^{\left((\ell+1)\delta+(\overline{\lambda}_1+\varepsilon+\delta)\right)(u-s)} \qquad \text{for } i = 1, \ldots, m+1-\ell.$$

Consequently, for any $t_2 > t_1 > s$ and $i = 1, \ldots, m+1-\ell$, we get

$$\left|\int_{t_1}^{t_2} \Lambda_1(s,u)\Lambda_2(u,s)^\ell \frac{d^i g_{u,s}}{\partial x_1^i}(x_1)\, du\right| \leq \widehat{C}_2 \frac{e^{(t_2-t_1)\alpha_2} - e^{(t_1-s)\alpha_2}}{\alpha_2},$$

where

$$\widehat{C}_2 := C_2 K(\varepsilon)^{\ell+1}, \quad \alpha_2 := \ell\overline{\lambda}_2 - \underline{\lambda}_1 + \overline{\lambda}_1 + (\ell+2)(\varepsilon+\delta).$$

Due to the fact that ε, δ satisfy the spectral gap condition (ii), the uniform convergence in x_1 in (2.75) is shown. Therefore, the function h_s is $C^{m+1-\ell}$.
Step 4: In this step, we show that for a.a. $s \in \mathbb{R}$ the map $\frac{\partial h_s}{\partial s}$ is $C^{m-\ell}$. By (2.71) and Step 1, it is sufficient to show that the following map

$$x_1 \mapsto \int_s^\infty \Lambda_1(s,u)\Lambda_2(u,s)^\ell \frac{\partial g_{u,s}}{\partial s}(x_1)\, du$$

is $C^{m-\ell}$. From (2.72), we derive that

$$\frac{\partial g_{u,s}}{\partial s}(x_1) = \frac{\exp\left(\ell \int_s^u q(v, \Phi_1(v,s,x_1))\, dv\right)}{\exp\left(\int_s^u \frac{\partial p}{\partial x_1}(v, \Phi_1(v,s,x_1))\, dv\right)} (T_1 + T_2),$$

where

$$T_1 = -\frac{\partial b}{\partial x_1}(u, \Phi_1(u,s,x_1)) \frac{\partial \Phi_1}{\partial x_1}(u,s,x_1)(a_1(s)x_1 + p(s,x_1)),$$

and

$$T_2 = b(u, \Phi_1(u,s,x_1)) \left(\frac{\partial p}{\partial x_1}(s,x_1) - \ell q(s,x_1) - (a_1(s)x_1 + p(s,x_1)) \cdot \right.$$
$$\left. \int_s^u \left(\ell \frac{\partial q}{\partial x_1}(v, \Phi_1(v,s,x_1)) - \frac{\partial^2 p}{\partial x_1^2}(v, \Phi_1(v,s,x_1))\right) \frac{\partial \Phi_1}{\partial x_1}(v,s,x_1)\, dv \right).$$

Analog to the proof in Step 3, we get that $\frac{\partial h_s}{\partial s}$ is $C^{m-\ell}$.
Step 5: Using the function h_s, we define the following map $\widetilde{\varphi} : \mathbb{R} \times \mathbb{R} \times \mathbb{R}^2 \to \mathbb{R}^2$ by

$$\widetilde{\varphi}(t, s, x_1, x_2) := H_t^{-1} \circ \varphi(t, s) \circ H_s(x_1, x_2), \tag{2.76}$$

where $H_t(x_1, x_2) := \begin{pmatrix} x_1 + h_t(x_1)x_2^\ell \\ x_2 \end{pmatrix}$. By the implicit function theorem and Steps 3 and 4, we derive that

$$H_t^{-1}(x_1, x_2) := \begin{pmatrix} x_1 + \alpha_t(x_1, x_2)x_2^\ell \\ x_2 \end{pmatrix}, \tag{2.77}$$

where α_t is $C^{m+1-\ell}$ and $\frac{d\alpha_t}{dt}$ is $C^{m-\ell}$. We show that the nonautonomous differential equation corresponding to $\widetilde{\varphi}$ is of the form of Eq. (2.70), and this completes the proof of this proposition. Indeed, by (2.76), we get that

$$\widetilde{\varphi}(t, s, \widetilde{\varphi}(s, \tau)(x_1, x_2)) = \widetilde{\varphi}(t, \tau)(x_1, x_2) \qquad \text{for } t, s, \tau \in \mathbb{R}. \tag{2.78}$$

Define

$$\widetilde{f}(t, x) := \frac{\partial \widetilde{\varphi}}{\partial t}(t, s, x)\Big|_{s=t} \qquad \text{for } x \in \mathbb{R}^2,$$

where the existence and smoothness of the function f are obtained from Step 4. By (2.78), we get

$$\begin{aligned} \frac{\partial \widetilde{\varphi}}{\partial t}(t, s, x) &= \lim_{\Delta \to 0} \frac{\widetilde{\varphi}(t + \Delta, s, x) - \widetilde{\varphi}(t, s, x)}{\Delta} \\ &= \lim_{\Delta \to 0} \frac{\widetilde{\varphi}(t + \Delta, t, \widetilde{\varphi}(t, s, x)) - \widetilde{\varphi}(t, s, x)}{\Delta} \\ &= \widetilde{f}(t, \widetilde{\varphi}(t, s, x)), \end{aligned}$$

which implies that $\widetilde{\varphi}(t, s, x)$ is the solution of the following nonautonomous equation

$$\dot{x} = \widetilde{f}(t, x).$$

Moreover, by definition of $\widetilde{\varphi}$ and (2.77), we get that

$$\widetilde{\varphi}(t, s, x) = \begin{pmatrix} \varphi_1(t, s, H_s(x)) + \alpha_t(\varphi_1(t, s, H_s(x)))\varphi_2(t, s, H_s(x))^\ell \\ \varphi_2(t, s, H_s(x)) \end{pmatrix},$$

which implies that $\widetilde{f}$ is in the class $C(m,n)$. Next, by definition of h as in Step 2, we get

$$h_t(x_1) = \lim_{\tau\to\infty}\lim_{x_2\to 0}\frac{\varphi_1(t,\tau)\circ\Phi(\tau,t)(x_1,x_2)-x_1}{x_2^\ell}.$$

Consequently, for any $s\in\mathbb{R}$, we have

$$\begin{aligned}
h_t(\Phi_1(t,s,x_1)) &= \lim_{\tau\to\infty}\lim_{x_2\to 0}\frac{\varphi_1(t,\tau)\circ\Phi(\tau,s)(x_1,x_2)-\Phi_1(t,s,x_1)}{\Phi_2(t,s,x_1,x_2)^\ell}\\
&= \lim_{\tau\to\infty}\lim_{x_2\to 0}\frac{\varphi_1(t,s)\circ\varphi(s,\tau)\circ\Phi(\tau,s)(x_1,x_2)-\Phi_1(t,s,x_1)}{\Phi_2(t,s,x_1,x_2)^\ell}\\
&= \lim_{\tau\to\infty}\lim_{x_2\to 0}\frac{\varphi_1(t,s,x_1+h_{s,\tau}(x_1)x_2^\ell,x_2)-\Phi_1(t,s,x_1)}{\Phi_2(t,s,x_1,x_2)^\ell}.
\end{aligned}$$

By Step 2, the function $h_{t,\tau}(x_1)$ converges uniformly exponentially to $h_t(x_1)$ as τ tends to infinity. Consequently, we get

$$h_t(\Phi_1(t,s,x_1)) = \lim_{x_2\to 0}\frac{\varphi_1(t,s,x_1+h_s(x_1)x_2^\ell,x_2)-\Phi_1(t,s,x_1)}{\Phi_2(t,s,x_1,x_2)^\ell},$$

which implies that

$$\begin{aligned}
&\Phi_1(t,s,x_1)+h_t(\Phi_1(t,s,x_1))\Phi_2(t,s,x_1,x_2)^\ell\\
&\quad= \varphi_1(t,s,x_1+h_s(x_1)x_2^\ell,x_2)+o(|x_2^\ell|).
\end{aligned}$$

Therefore, by definition of $\widetilde{\varphi}$, we get

$$\widetilde{\varphi}_1(t,s,x_1,x_2) = \Phi_1(t,s,x_1)+o(x_2^\ell),$$

which implies that the function $\frac{\partial\widetilde{\varphi}_1}{\partial t}(t,t,x)$ is different from $\frac{\partial\Phi_1}{\partial t}(t,t,x)$ a term of order $x_2^{\ell+1}$. Note that

$$\frac{\partial\widetilde{\varphi}_1}{\partial t}(t,t,x) = \widetilde{f}_1(t,x), \qquad \frac{\partial\Phi_1}{\partial t}(t,t,x) = a_1(t)x_1+p(t,x_1),$$

which completes the proof.

Analog to Proposition 2.4, we also have the following proposition. Note that the structure of the proof of Proposition 2.5 is quite similar to the proof of Proposition 2.4, but only the terms being removed are different. Hence, we will only formulate the result in each step without details of the proof.

Proposition 2.5 *Let $\ell \in \{1, \ldots, n-2\}$ and $m \geq n$ be arbitrary. Consider the following system*

$$\dot{x} = \begin{pmatrix} a_1(t) & 0 \\ 0 & a_2(t) \end{pmatrix} x + \begin{pmatrix} p(t, x_1) \\ q(t, x_1)x_2 \end{pmatrix} + \begin{pmatrix} b(t, x_1)x_2^{\ell+1} \\ c(t, x_1)x_2^{\ell+1} \end{pmatrix} + r(t, x), \qquad (2.79)$$

where $b(t, x_1), c(t, x_1)$ are $C^{m-\ell}$ Carathéodory functions and $r(t, x)$ is in class $C(m+1, n)$ such that $r(t, x) = O(x_2^{\ell+2})$. Suppose that the following spectral gap condition holds:

(i) If $\overline{\lambda}_1 \geq 0$, then $\ell\overline{\lambda}_2 + (m-\ell)\overline{\lambda}_1 + m(\varepsilon + \delta) < 0$.
(ii) If $\overline{\lambda}_1 < 0$, then $\ell\overline{\lambda}_2 + \overline{\lambda}_1 + (\ell+1)(\varepsilon + \delta) < 0$.

Then, system (2.69) is $C^{m-\ell}$ equivalent to a system of the following form

$$\dot{y} = \begin{pmatrix} a_1(t) & 0 \\ 0 & a_2(t) \end{pmatrix} y + \begin{pmatrix} p(t, y_1) \\ q(t, y_1)y_2 \end{pmatrix} + w(t, y), \qquad (2.80)$$

where $w(t, y)$ is in class $C(m, n)$ with $w_1(t, y) = O(|y_2|^{\ell+1})$ and $w_2(t, y) = O(|y_2|^{\ell+2})$.

Proof Let $\varphi(t, s, x_1, x_2)$ denote the solution of (2.79) with the initial condition $x(s) = (x_1, x_2)^{\mathrm{T}}$. Recall that $\Lambda(t, s)x$, resp. $\Phi(t, s, x)$, denote the solution satisfying the initial value problem $x(s) = x$ of the following systems

$$\dot{x} = \begin{pmatrix} a_1(t) & 0 \\ 0 & a_2(t) \end{pmatrix} x, \quad \text{resp.} \quad \dot{x} = \begin{pmatrix} a_1(t) & 0 \\ 0 & a_2(t) \end{pmatrix} x + \begin{pmatrix} p(t, x_1) \\ q(t, x_1)x_2 \end{pmatrix}.$$

In order to construct a smooth transformation between system (2.79) and system (2.80), we define a scalar function $h_{s,t} : \mathbb{R} \to \mathbb{R}$, $s, t \in \mathbb{R}$, by

$$h_{s,t}(x_1) := \lim_{x_2 \to 0} \frac{\varphi_2(s, t) \circ \Phi(t, s)(x_1, x_2) - x_2}{x_2^{\ell+1}} \qquad \text{for } x_1 \in \mathbb{R}.$$

In the following steps, we investigate some properties of the function $h_{s,t}$ and explain how to construct the desired transformation by using the function $h_{s,t}$.
Step 1: In this step, we prove that

$$h_{s,t}(x_1) = -\int_s^t \Lambda_2(u, s)^\ell g_{u,s}(x_1)\, du, \qquad (2.81)$$

where

$$g_{u,s}(x_1) := b(u, \Phi_1(u, s, x_1)) \exp\left(\ell \int_s^u q(v, \Phi_1(v, s, x_1))\, dv\right). \qquad (2.82)$$

Step 2: In this step, we show the existence of the limit function $h_s : \mathbb{R} \to \mathbb{R}$ defined by $h_s(x_1) = \lim_{t\to\infty} h_{s,t}(x_1)$ by arguing that for $t_2 > t_1 > s$

$$|h_{s,t_2}(x_1) - h_{s,t_1}(x_1)| \le MK(\varepsilon)^\ell \int_{t_1}^{t_2} e^{\ell(\overline{\lambda}_2+\varepsilon+\delta)(u-s)}\, du.$$

Since $\overline{\lambda}_2 + \varepsilon + \delta < 0$, it follows that the limit $\lim_{t\to\infty} h_{s,t}(x_1)$ exists uniformly in x_1.

Step 3: In this step, we show that the function h_s is $C^{m-\ell}$. It is sufficient to show the uniform convergence in x_1 of the following improper integral

$$\int_s^\infty \Lambda_2(u,s)^\ell \frac{d^i g_{u,s}}{dx_1^i}(x_1)\, du \qquad \text{for } i = 1, \dots, m-\ell. \tag{2.83}$$

For this purpose, we consider the following two cases $\overline{\lambda}_1 \ge 0$ and $\overline{\lambda}_1 < 0$.

Case 1: $\overline{\lambda}_1 \ge 0$. Using Lemma 2.8 and Theorem 2.7, it is seen that there is a positive constant C_1 such that for $u \ge s$

$$\left|\frac{d^i g_{u,s}}{\partial x_1^i}(x_1)\right| \le C_1 e^{\left(\ell\delta + i(\overline{\lambda}_1+\varepsilon+\delta)\right)(u-s)} \qquad \text{for } i = 1, \dots, m-\ell.$$

Consequently, for any $t_2 > t_1 > s$ and $i = 1, \dots, m-\ell$, we get

$$\left|\int_{t_1}^{t_2} \Lambda_2(u,s)^\ell \frac{d^i g_{u,s}}{\partial x_1^i}(x_1)\, du\right| \le \widehat{C}_1 \frac{e^{(t_2-s)\alpha_1} - e^{(t_1-s)\alpha_1}}{\alpha_1},$$

where

$$\widehat{C}_1 := C_1 K(\varepsilon)^\ell, \quad \alpha_1 := \ell\overline{\lambda}_2 + i\overline{\lambda}_1 + (\ell+i)(\varepsilon+\delta),$$

which proves the uniform convergence in x_1 in (2.83). Therefore, the function h_s is $C^{m-\ell}$.

Case 2: $\overline{\lambda}_1 < 0$. Using Lemma 2.8 and Theorem 2.7, it is seen that there is a positive constant C_2 such that for $u \ge s$

$$\left|\frac{d^i g_{u,s}}{\partial x_1^i}(x_1)\right| \le C_2 e^{\left(\ell\delta + (\overline{\lambda}_1+\varepsilon+\delta)\right)(u-s)} \qquad \text{for } i = 1, \dots, m-\ell.$$

Consequently, for $t_2 > t_1 > s$ and $i = 1, \dots, m-\ell$, we get

$$\left|\int_{t_1}^{t_2} \Lambda_2(u,s)^\ell \frac{d^i g_{u,s}}{\partial x_1^i}(x_1)\, du\right| \le \widehat{C}_2 \frac{e^{(t_2-s)\alpha_2} - e^{(t_1-s)\alpha_2}}{\alpha_2},$$

where

$$\widehat{C}_2 := C_2 K(\varepsilon)^{\ell+1}, \quad \alpha_2 := \ell\overline{\lambda}_2 + \overline{\lambda}_1 + (\ell+1)(\varepsilon+\delta),$$

which proves uniform convergence in x_1 in (2.83). Therefore, the function h_s is $C^{m-\ell}$.

Step 4: In this step, we show that for a.a. $s \in \mathbb{R}$, the function $\frac{\partial h_s}{\partial s}$ is $C^{m-\ell-1}$. By (2.81) and Step 3, it is sufficient to show that the following map

$$x_1 \mapsto \int_s^\infty \Lambda_2(u,s)^\ell \frac{\partial g_{u,s}}{\partial s}(x_1)\, du$$

is $C^{m-\ell-1}$. From (2.82), we derive that

$$\frac{\partial g_{u,s}}{\partial s}(x_1) = -\exp\Big(\ell \int_s^u q(v, \Phi_1(v,s,x_1))\, dv\Big)(T_1 + T_2),$$

where

$$T_1 = \frac{\partial b}{\partial x_1}(u, \Phi_1(u,s,x_1)) \frac{\partial \Phi_1}{\partial x_1}(u,s,x_1)\,(a_1(s)x_1 + p(s,x_1))$$

and

$$T_2 = \ell\, b(u, \Phi_1(u,s,x_1))\Big(q(s,x_1) + (a_1(s)x_1 + p(s,x_1)) \cdot \int_s^u \frac{\partial q}{\partial x_1}(v, \Phi_1(v,s,x_1)) \frac{\partial \Phi_1}{\partial x_1}(v,s,x_1)\, dv\Big).$$

Analog to the proof in Step 3, we get that $\frac{\partial h_s}{\partial s}$ is $C^{m-\ell-1}$.

Step 5: Using the function h_s, we define the following function $\widetilde{\varphi} : \mathbb{R}\times\mathbb{R}\times\mathbb{R}^2 \to \mathbb{R}^2$ by

$$\widetilde{\varphi}(t,s,x_1,x_2) := H_t^{-1} \circ \varphi(t,s) \circ H_s(x_1,x_2),$$

where $H_t(x_1,x_2) := \begin{pmatrix} x_1 \\ x_2 + h_t(x_1)x_2^{\ell+1} \end{pmatrix}$. Similar to Step 5 in the proof of Proposition 2.4, one can show that the nonautonomous differential equation corresponding to $\widetilde{\varphi}$ is of the form of Eq. (2.80), and this completes the proof of this proposition.

We are now at a position to state and prove the main result of this subsection.

Theorem 2.8 (Normalization Up to Finite Order Along Invariant Manifold) *Consider system* (2.60) *with C^N Carathéodory vector field. Let $n \in \mathbb{N}$ satisfy that $4n - 3 \le N$. Assume that the following spectral gap condition holds:*

$$\lambda_2 + (4n-4)\lambda_1 < \lambda_1.$$

Then, system (2.60) *is C^n equivalent to a system of the form*

$$\dot{x} = \begin{pmatrix} a_1(t) & 0 \\ 0 & a_2(t) \end{pmatrix} x + \begin{pmatrix} p(t,x_1) \\ q(t,x_1)x_2 \end{pmatrix} + w(t,x),$$

where w is a C^n Carathéodory function and $w(t,x) = O(x_2^n)$.

Proof We expand (2.60) into the following form

$$\dot{x} = \begin{pmatrix} a_1(t) & 0 \\ 0 & a_2(t) \end{pmatrix} x + \begin{pmatrix} p(t,x_1) \\ q(t,x_1)x_2 \end{pmatrix} + \begin{pmatrix} \sum_{i=1}^{n-1} \frac{1}{i!}\frac{\partial^i r_1}{\partial x_2^i}(t,x_1,0)x_2^i \\ \sum_{i=2}^{n-1} \frac{1}{i!}\frac{\partial^i r_2}{\partial x_2^i}(t,x_1,0)x_2^i \end{pmatrix}$$
$$+ O(|x_2|^n).$$

Applying Propositions 2.4 and 2.5, we can eliminate step by step the following terms

$$\sum_{i=1}^{n-1} \frac{1}{i!}\frac{\partial^i r_1}{\partial x_2^i}(t,x_1,0)x_2^i \quad \text{and} \quad \sum_{i=2}^{n-1} \frac{1}{i!}\frac{\partial^i r_2}{\partial x_2^i}(t,x_1,0)x_2^i$$

by transformations which is quasi-homogeneous with respect to the variable x_2 and does not affect lower-order terms. Note that in each elimination step, we loose one differentiability order of the vector field. Hence, by starting with a vector field r in class $C(4n-3,n)$ and eliminating $2n-3$ terms, we arrive at a vector field such that the remaining term w is in class $C(2n,n)$, and thus we get that w is a C^n Carathéodory function. The proof is complete.

2.3.5 *Elimination of the Remaining Term*

According to Theorem 2.8, under the spectral gap condition as in Theorem 2.6, system (2.60) is C^n equivalent to a system of the following form

$$\dot{x} = \begin{pmatrix} a_1(t) & 0 \\ 0 & a_2(t) \end{pmatrix} x + \begin{pmatrix} p(t,x_1) \\ q(t,x_1)x_2 \end{pmatrix} + r(t,x), \tag{2.84}$$

where r is a C^n Carathéodory function with $r(t,x) = O(|x_2|^n)$. Therefore, to prove Theorem 2.6, it remains to show that (2.84) is C^k equivalent to the partially linear system, which is recalled for convenience, as follows

$$\dot{x} = \begin{pmatrix} a_1(t) & 0 \\ 0 & a_2(t) \end{pmatrix} x + \begin{pmatrix} p(t, x_1) \\ q(t, x_1)x_2 \end{pmatrix}. \tag{2.85}$$

Before doing this, we can assume by using the cutoff technique (Example 2.3) that there exists a constant C such that

$$\left\| \frac{\partial^{i+j} r}{\partial x_1^i \partial x_2^j}(t, x) \right\| \le C|x_2|^{n-j} \qquad \text{for } 0 \le i + j \le n. \tag{2.86}$$

By the spectral gap condition in Theorem 2.6, we can choose and fix constants ε, δ such that

Case 1: If $\overline{\lambda}_1 \ge 0$, then

$$n\overline{\lambda}_2 - k\underline{\lambda}_2 + k\overline{\lambda}_1 + (n + 2k)(\varepsilon + 2\delta) < 0. \tag{2.87}$$

Case 2: If $\overline{\lambda}_1 < 0$, then $\overline{\lambda}_1 + \varepsilon + \delta < 0$ and

$$n\overline{\lambda}_2 - \underline{\lambda}_2 + (n + 1)(\varepsilon + \delta), \quad n\overline{\lambda}_2 - k\underline{\lambda}_2 + \overline{\lambda}_1 + (n + 2k)(\varepsilon + 2\delta) < 0. \tag{2.88}$$

Using the cutoff function technique again, we can assume that the function $\widehat{r} : \mathbb{R} \times \mathbb{R}^2 \to \mathbb{R}^2$ defined by

$$\widehat{r}(t, x) := \begin{pmatrix} p(t, x_1) \\ q(t, x_1)x_2 \end{pmatrix} + r(t, x),$$

satisfies the assumption of Lemma 2.6, i.e., $\sup_{(t,x)\in\mathbb{R}\times\mathbb{R}^2} \|D_x\widehat{r}(t, x)\| \le \frac{\delta}{2K(\varepsilon)}$. We are now in a position to prove our main result Theorem 2.6.

Proof of Theorem 2.6 Let $\varphi(t, s, x)$ denote the solution of (2.84) satisfying $x(s) = x$. Recall that $\Lambda(t, s)x$, resp. $\Phi(t, s, x)$, denote the solutions satisfying the initial value problems $x(s) = x$ of the following systems

$$\dot{x} = \begin{pmatrix} a_1(t) & 0 \\ 0 & a_2(t) \end{pmatrix} x, \quad \text{resp.} \quad \dot{x} = \begin{pmatrix} a_1(t) & 0 \\ 0 & a_2(t) \end{pmatrix} x + \begin{pmatrix} p(t, x_1) \\ q(t, x_1)x_2 \end{pmatrix}.$$

In order to construct a smooth transformation between system (2.84) and system (2.85), we define the following function $H_s(t, \cdot) : \mathbb{R}^2 \to \mathbb{R}^2$, where $s, t \in \mathbb{R}$, by

$$H_s(t, x) := \varphi(s, t, \Phi(t, s, x)) \qquad \text{for } x \in \mathbb{R}^2. \tag{2.89}$$

In the following steps, we investigate some properties of the function $H_s(t, \cdot)$ and explain how to construct the desired transformation by using the function $H_s(t, \cdot)$.

Step 1: In this step, we show that

$$H_s(t,x) = x - \int_s^t \frac{\partial \varphi}{\partial x}(s,u,\Phi(u,s,x))r(u,\Phi(u,s,x))\,du. \tag{2.90}$$

Thereto, taking the derivative with respect to the t-variable in (2.89) leads to

$$\frac{\partial H_s}{\partial t}(t,x) = \frac{\partial \varphi}{\partial t}(s,t,\Phi(t,s,x)) + \frac{\partial \varphi}{\partial x}(s,t,\Phi(t,s,x))\frac{d}{dt}\Phi(t,s,x). \tag{2.91}$$

By definition of Φ, we get that

$$\frac{d}{dt}\Phi(t,s,x) = \begin{pmatrix} a_1(t) & 0 \\ 0 & a_2(t) \end{pmatrix} \Phi(t,s,x) + \begin{pmatrix} p(t,\Phi_1(t,s,x)) \\ q(t,\Phi_1(t,s,x))\Phi_2(t,s,x) \end{pmatrix}. \tag{2.92}$$

Therefore, in order to get a more explicit form of (2.91), it remains to deal with the term $\frac{\partial \varphi}{\partial t}(s,t,\Phi(t,s,x))$. Note that for any $z \in \mathbb{R}^2$

$$\frac{\partial \varphi}{\partial t}(s,t,z) = \lim_{\Delta \to 0} \frac{\varphi(s,t+\Delta,z) - \varphi(s,t,z)}{\Delta},$$

which together with the evolution property implies that

$$\begin{aligned} \frac{\partial \varphi}{\partial t}(s,t,z) &= \lim_{\Delta \to 0} \frac{\varphi(s,t,\varphi(t,t+\Delta,z)) - \varphi(s,t,z)}{\Delta} \\ &= \frac{\partial \varphi}{\partial x}(s,t,z) \lim_{\Delta \to 0} \frac{\varphi(t,t+\Delta,z) - z}{\Delta}. \end{aligned} \tag{2.93}$$

Since

$$\varphi(t,t+\Delta,z) = z - \Delta \left(\begin{pmatrix} a_1(t) & 0 \\ 0 & a_2(t) \end{pmatrix} z + \begin{pmatrix} p(t,z_1) \\ q(t,z_1)z_2 \end{pmatrix} + r(t,z) \right) + O(|\Delta|^2),$$

it follows that

$$\frac{\partial \varphi}{\partial t}(t,t,z) = -\begin{pmatrix} a_1(t) & 0 \\ 0 & a_2(t) \end{pmatrix} z - \begin{pmatrix} p(t,z_1) \\ q(t,z_1)z_2 \end{pmatrix} - r(t,z).$$

Replacing z by $\Phi(t,s,x)$, we obtain that

$$\frac{\partial \varphi}{\partial t}(t, t, \Phi(t, s, x)) = -\begin{pmatrix} a_1(t) & 0 \\ 0 & a_2(t) \end{pmatrix} \Phi(t, s, x) - \begin{pmatrix} p(t, \Phi_1(t, s, x)) \\ q(t, \Phi_1)\Phi_2(t, s, x) \end{pmatrix}$$

$$-r(t, \Phi(t, s, x)),$$

which together with (2.93) gives that

$$\frac{\partial \varphi}{\partial t}(s, t, \Phi(t, s, x)) = -\frac{\partial \varphi}{\partial x}(s, t, \Phi(t, s, x)) \left(\begin{pmatrix} a_1(t) & 0 \\ 0 & a_2(t) \end{pmatrix} \Phi(t, s, x) + \begin{pmatrix} p(t, \Phi_1(t, s, x)) \\ q(t, \Phi_1)\Phi_2(t, s, x) \end{pmatrix} + r(t, \Phi(t, s, x)) \right).$$

Combining the above equality and (2.91) and (2.92), we obtain that

$$\frac{\partial H_s}{\partial t}(t, x) = -\frac{\partial \varphi}{\partial x}(s, t, \Phi(t, s, x))r(t, \Phi(t, s, x)),$$

which together with the fact that $H_s(s, x) = x$ proves (2.90).
***Step 2*:** In this step, we show that the map

$$H_s(x) := \lim_{t \to \infty} H_s(t, x) = x - \int_s^\infty \frac{\partial \varphi}{\partial x}(s, u, \Phi(u, s, x))r(u, \Phi(u, s, x))\, du \tag{2.94}$$

is well-defined and fulfills the following properties:

(i) $\lim_{x \to 0} H_t(x) = 0$ uniformly in t.
(ii) H maps a solution of (2.85) to a solution of (2.84).

For this purpose, let $t_1 > t_2 \geq s$ be arbitrary. Using Lemma 2.9 and Lemma 2.6, we obtain that for some positive constant L

$$\begin{aligned} &\left\| \int_{t_2}^{t_1} \frac{\partial \varphi}{\partial x}(s, u, \Phi(u, s, x))r(u, \Phi(u, s, x))\, du \right\| \\ &\leq L \int_{t_2}^{t_1} e^{(u-s)(\varepsilon+\delta-\underline{\lambda}_2)} \|r(u, \Phi(u, s, x))\|\, du \\ &\leq CLK(\varepsilon)^n |x_2|^n \int_{t_2}^{t_1} e^{(u-s)\left(n\overline{\lambda}_2 - \underline{\lambda}_2 + (n+1)(\varepsilon+\delta)\right)}\, du. \end{aligned}$$

By (2.87) and (2.88), we get that $n\overline{\lambda}_2 - \underline{\lambda}_2 + (n+1)(\varepsilon+\delta) < 0$. Thus, the following improper integral

$$\int_s^\infty \frac{\partial \varphi}{\partial x}(s, u, \Phi(u, s, x))r(u, \Phi(u, s, x))\,du$$

exists. Furthermore, we have

$$\begin{aligned}\|H_s(x) - x\| &\le CLK(\varepsilon)^n |x_2|^n \int_s^\infty e^{(u-s)\left(n\overline{\lambda}_2 - \underline{\lambda}_2 + (n+1)(\varepsilon+\delta)\right)}\,du \\ &= \frac{CLK(\varepsilon)^n}{n\overline{\lambda}_2 - \underline{\lambda}_2 + (n+1)(\varepsilon+\delta)}|x_2|^n,\end{aligned}$$

which proves (i). Thus, it remains to prove that H maps a solution of (2.85) to a solution of (2.84). By definition of H, we get that for $t, s \in \mathbb{R}$ and $x \in \mathbb{R}^2$

$$\begin{aligned}H_t(\Phi(t, s, x)) &= \lim_{u\to\infty} \varphi(t, u, \Phi(u, t) \circ \Phi(t, s, x)) \\ &= \varphi(t, s, \lim_{u\to\infty} \varphi(s, u, \Phi(u, s, x))) \\ &= \varphi(t, s, H_s(x)),\end{aligned}$$

which proves that H maps a solution of (2.85) to a solution of (2.84).

Step 3: In this step, we show that H_s is C^k. By definition of H in (2.94), it is sufficient to show for $1 \le i + j := m \le k$ the uniform convergence in x of the following improper integrals

$$\int_s^\infty \frac{\partial^{i+j}}{\partial x_1^i \partial x_2^j}\frac{\partial \varphi_\alpha}{\partial x_\beta}(s, u, \Phi(u, s, x))r_\beta(u, \Phi(u, s, x))\,du \qquad \text{for } \alpha, \beta \in \{1, 2\}. \tag{2.95}$$

Using the generalized Leibniz's product rule, see, e.g., [70, Proposition 6]; we get

$$\frac{\partial^{i+j}}{\partial x_1^i \partial x_2^j}\frac{\partial \varphi_\alpha}{\partial x_\beta}(s, u, \Phi(u, s, x))r_\beta(u, \Phi(u, s, x))$$

$$\sum_{\widehat{i}=0}^{i}\sum_{\widehat{j}=0}^{j}\left(\frac{\partial^{\widehat{i}+\widehat{j}}}{\partial x_1^{\widehat{i}}\partial x_2^{\widehat{j}}}\frac{\partial \varphi_\alpha}{\partial x_\beta}(s, u, \Phi(u, s, x))\right)\left(\frac{\partial^{(i+j)-(\widehat{i}+\widehat{j})}}{\partial x_1^{i-\widehat{i}}\partial x_2^{j-\widehat{j}}}r_\beta(u, \Phi(u, s, x))\right).$$

It therefore remains to estimate the following product

$$\left(\frac{\partial^{\widehat{i}+\widehat{j}}}{\partial x_1^{\widehat{i}}\partial x_2^{\widehat{j}}}\frac{\partial \varphi_\alpha}{\partial x_\beta}(s, u, \Phi(u, s, x))\right)\left(\frac{\partial^{(i+j)-(\widehat{i}+\widehat{j})}}{\partial x_1^{i-\widehat{i}}\partial x_2^{j-\widehat{j}}}r_\beta(u, \Phi(u, s, x))\right).$$

For this purpose, we consider the following two cases.

Case 1: $\overline{\lambda}_1 \geq 0$.

According to Lemma 2.8, Lemma 2.9, inequality (2.86) and Lemma 2.6, it is easily seen that there exists a constant C_1 such that for any $\widehat{i} \leq i$ and $\widehat{j} \leq j$ and $u \geq s$, we have

$$\left|\left(\frac{\partial^{\widehat{i}+\widehat{j}}}{\partial x_1^{\widehat{i}} \partial x_2^{\widehat{j}}} \frac{\partial \varphi_\alpha}{\partial x_\beta}(s, u, \Phi(u, s, x))\right)\left(\frac{\partial^{(i+j)-(\widehat{i}+\widehat{j})}}{\partial x_1^{i-\widehat{i}} \partial x_2^{j-\widehat{j}}} r_\beta(u, \Phi(u, s, x))\right)\right|$$
$$\leq C_1 e^{(n\overline{\lambda}_2 - k\underline{\lambda}_2 + k\overline{\lambda}_1 + (n+2k)(\varepsilon+2\delta))(u-s)} |x_2|^{n-j+\widehat{j}}.$$

From (2.87), we derive that $n\overline{\lambda}_2 - k\underline{\lambda}_2 + k\overline{\lambda}_1 + (n+2k)(\varepsilon+2\delta) < 0$ and therefore the uniform convergence in x of the improper integral (2.95) is proved.

Case 2: $\overline{\lambda}_1 < 0$.

According to Lemma 2.8, Lemma 2.9, inequality (2.86) and Lemma 2.6, it is easily seen that there exists a constant C_2 such that for any $\widehat{i} \leq i$ and $\widehat{j} \leq j$ and $u \geq s$, we have

$$\left|\left(\frac{\partial^{\widehat{i}+\widehat{j}}}{\partial x_1^{\widehat{i}} \partial x_2^{\widehat{j}}} \frac{\partial \varphi_\alpha}{\partial x_\beta}(s, u, \Phi(u, s, x))\right)\left(\frac{\partial^{(i+j)-(\widehat{i}+\widehat{j})}}{\partial x_1^{i-\widehat{i}} \partial x_2^{j-\widehat{j}}} r_\beta(u, \Phi(u, s, x))\right)\right|$$
$$\leq C_2 e^{(n\overline{\lambda}_2 - k\underline{\lambda}_2 + \overline{\lambda}_1 + (n+2k)(\varepsilon+2\delta))(u-s)} |x_2|^{n-j+\widehat{j}}.$$

By (2.88), we get that $n\overline{\lambda}_2 - k\underline{\lambda}_2 + \overline{\lambda}_1 + (n+2k)(\varepsilon+2\delta) < 0$ and thus the uniform convergence in x of the improper integral (2.95) is shown. The proof is complete.

Remark 2.3 The statement of Theorem 2.6 also holds under the dual assumption $\lambda_1 < \lambda_2$, $\lambda_2 > 0$, and there exists $n \in \mathbb{N}$ such that $4n - 3 < N$ and

(B1) If $\underline{\lambda}_1 \leq 0$, then $\lambda_2 + (4n-4)\lambda_1 > \lambda_1$ and $n\lambda_2 + k\lambda_1 - k\lambda_2 > 0$.

(B2) If $\underline{\lambda}_1 > 0$, then $n\lambda_2 - \lambda_2 > 0$ and $n\lambda_2 + \lambda_1 - k\lambda_2 > 0$.

The proof is similar to the proof of Theorem 2.6 by using the time reversal.

2.3.6 *Example*

Consider a *Duffing-van der Pol oscillator* under the influence of a nonautonomous parametric perturbation

$$\ddot{y} = (\alpha + \sigma\xi(t))y + \beta y - y^3 - y^2\dot{y}, \tag{2.96}$$

where $\xi : \mathbb{R} \to [-1, 1]$ is a bounded measurable function and σ is an intensity parameter. We refer the reader to [6, Chapters 8, 9] and [111] for a motivation to investigate this equation. With $x = (x_1, x_2)^{\mathrm{T}} = (y, \dot{y})^{\mathrm{T}}$, system (2.96) becomes

$$\dot{x} = \begin{pmatrix} 0 & 1 \\ 0 & \beta \end{pmatrix} x + (\alpha + \sigma\xi(t)) \begin{pmatrix} 0 & 0 \\ 1 & 0 \end{pmatrix} x + \begin{pmatrix} 0 \\ -x_1^3 - x_1^2 x_2 \end{pmatrix}.$$

We choose $\beta = -1$ and treat (α, σ) as a small two-dimensional parameter. The preceding equation can be rewritten in the new coordinates $T^{-1}x$ with $T = \begin{pmatrix} 1 & -1 \\ 0 & 1 \end{pmatrix}$ as follows

$$\dot{x} = \begin{pmatrix} \alpha + \sigma\xi(t) & -\alpha - \sigma\xi(t) \\ \alpha + \sigma\xi(t) & -1 - \alpha - \sigma\xi(t) \end{pmatrix} x - \begin{pmatrix} x_1^3 - 2x_1^2 x_2 + x_1 x_2^2 \\ x_1^3 - 2x_1^2 x_2 + x_1 x_2^2 \end{pmatrix}. \tag{2.97}$$

The linearization of (2.97) at $x = 0$ is

$$\dot{v} = \begin{pmatrix} \alpha + \sigma\xi(t) & -\alpha - \sigma\xi(t) \\ \alpha + \sigma\xi(t) & -1 - \alpha - \sigma\xi(t) \end{pmatrix} v.$$

Using the diagonal dominance criterion for exponential dichotomy, see, e.g., [98]; we obtain that for sufficiently small α, σ, the dichotomy spectrum of the above linearization consists of two disjoint intervals $\Sigma = \lambda_1 \cup \lambda_2$ and

$$\lambda_1 \subset \big[-2|\alpha| - 2|\sigma|, 2|\alpha| + 2|\sigma| \big],$$

$$\lambda_2 \subset \big[-1 - 2|\alpha| - 2|\sigma|, -1 + 2|\alpha| + 2|\sigma| \big].$$

Note that for sufficiently small α and σ, there exist k, N such that the spectral gap conditions in Theorems 2.5 and 2.6 are satisfied. Consequently, Eq. (2.97) is C^k equivalent to a system of the following form

$$\dot{x} = \Lambda(t)x + \begin{pmatrix} p(t, x_1) \\ q(t, x_1)x_2 \end{pmatrix},$$

where $p, q : R \times \mathbb{R} \to \mathbb{R}$ are C^k Carathéodory functions which satisfy

$$p(t, 0) = \frac{\partial p}{\partial x_1}(t, 0) = q(t, 0) = 0.$$

It remains to compute or numerically approximate the functions p and q. Note that after flattening the invariant manifold, we can compute explicitly the partial linearization form; see Sect. 2.3.1. Therefore, this task is equivalent to the computation of the invariant manifold. We refer the reader to [103] for a method to approximate the invariant manifold.

Chapter 3
Spectral Theory for Random Dynamical Systems

Abstract In this chapter, we introduce two spectral theories for linear random dynamical systems. The first one is the Lyapunov spectrum which is also known as the multiplicative ergodic theorem. The second one is the dichotomy spectrum (also known as Sacker-Sell spectrum) defined in terms of exponential dichotomy. We also establish explicit formulas for Lyapunov exponent of planar bilinear stochastic differential equations. The materials of this chapter are taken from Anh and Doan (Stoch Analy Appl 35(4):662–676, 2017) and Callaway et al. (Annales de l'Institut Henri Poincaré, Probabilités et Statistiques 53:1548–1574, 2017).

3.1 Basic Notions of Random Dynamical Systems

Roughly speaking, a random dynamical system consists of two ingredients. The first one is randomness which is modeled by an adequate metric dynamical system. The second one is dynamical rule which is reflected inside a suitable notion of cocycle. Our aim in this subsection is step by step to recall these fundamental notions.

3.1.1 Metric Dynamical Systems

Definition 3.1 A *metric dynamical system* with time $\mathbb{T}$, where $\mathbb{T} = \mathbb{R}$ or $\mathbb{Z}$ denoted by $(\Omega, \mathcal{F}, \mathbb{P}, (\theta_t)_{t\in\mathbb{T}})$,[1] is a probability space $(\Omega, \mathcal{F}, \mathbb{P})$ with a family of transformations $\theta_t : \Omega \to \Omega, t \in \mathbb{T}$ such that

(i) it is an one-parameter group, i.e.,

$$\theta_0 = \mathrm{id}_\Omega, \quad \theta_t \circ \theta_s = \theta_{t+s} \quad \text{for all } t, s \in \mathbb{T},$$

[1] In the case that $\mathbb{T}$ is discrete, i.e., $\mathbb{T} = \mathbb{Z}$, we use the notation $(\Omega, \mathcal{F}, \mathbb{P}, (\theta^n)_{n\in\mathbb{Z}})$ instead of the notation $(\Omega, \mathcal{F}, \mathbb{P}, (\theta_t)_{t\in\mathbb{R}})$.

T. S. Doan, *Spectral Theory of Nonautonomous Dynamical Systems and Applications*, https://doi.org/10.1007/978-981-97-5520-2_3

where id_Ω is the identical map on Ω,

(ii) The mapping $(t, \omega) \mapsto \theta_t\omega$ is $\mathcal{B}(\mathbb{T}) \otimes \mathcal{F}, \mathcal{F}$-measurable,

(iii) $\theta_t\mathbb{P} = \mathbb{P}$ for all $t \in \mathbb{T}$, i.e., $\mathbb{P}(\theta_t B) = \mathbb{P}(B)$ for all $B \in \mathcal{F}$ and $t \in \mathbb{T}$.

A set $B \in \mathcal{F}$ is called *invariant* if $\theta_t B = B$ for all $t \in \mathbb{T}$. A metric dynamical system θ is said to be *ergodic* under $\mathbb{P}$ if for any invariant set $B \in \mathcal{F}$ we have either $\mathbb{P}(B) = 0$ or $\mathbb{P}(B) = 1$.

Example 3.1 Examples of metric dynamical systems include the rotation map/ flow on the torus with irrationally frequency, sample path spaces of Brownian motion, and fractional Brownian motion equipped with the Wiener measures. We refer the readers to [6, 59, 66] for a more detail.

In what follows, we recall the representation theorem of an ergodic flow established in [2] which points out that any ergodic flow is isomorphic to a flow built under a function. To state this result, we need the following notion.

Definition 3.2 (Flow Built Under a Function) Let $(X, \mathcal{A}, m)$ be a probability space and $T : X \to X$ be an invertible measurable transformation preserving the probability m. Let $f : X \to \mathbb{R}_{\geq 0}$ be a measurable function with $\int_X f(x)\, dm(x) < \infty$. The flow build under the function (X, T, m, f) is defined with the following ingredients:

(i) *The base space* (B, σ, μ): The set B is defined by $B := \{(x, s) \in X \times \mathbb{R}_{\geq 0} : x \in X, 0 \leq s < f(x)\}$. Then, B is a measurable set of the measurable space $(X \times \mathbb{R}, \mathcal{A} \otimes \mathcal{B}(\mathbb{R}))$. Let σ and μ denote the restriction of the sigma algebra $\mathcal{A} \otimes \mathcal{B}(\mathbb{R})$, the measure $m \times \lambda$ on the measurable set B, respectively.

(ii) *The flow* $(S_t)_{t\in\mathbb{R}}$: For each $t \in \mathbb{R}_{\geq 0}$, the map $S_t : B \to B$ is defined by

$$S_t(x, s) := \left(T^{k-1}x, t + s - \sum_{j=1}^{k-1} f(T^j x) \right),$$

where k is the smallest positive integer satisfying

$$\sum_{j=1}^{k-1} f(T^j x) \leq t + s < \sum_{j=1}^{k} f(T^j x),$$

and for each $t < 0$, $S_t := (S_{-t})^{-1}$.

Theorem 3.1 (Representation of Ergodic Flow) *Suppose that $(\Omega, \mathcal{F}, \mathbb{P})$ is a Lebesgue space and $(\theta_t)_{t\in\mathbb{R}}$ be an ergodic flow from Ω into itself without fixed point preserving the probability $\mathbb{P}$. Then, there is a flow built under a function (X, T, m, f), where $(X, \mathcal{A}, m)$ is a Lebesgue space, T is an invertible ergodic transformation on X preserving the probability m, and the function f is bounded strictly away 0 and ∞, which is isomorphic to $(\theta_t)_{t\in\mathbb{R}}$.*

Proof From [2, Theorem 2] (see also [40, Chapter 11]), any ergodic flow is isomorphic to a flow built under a function (X, T, m, f). Furthermore, since $(\theta_t)_{t\in\mathbb{R}}$ is ergodic, it follows that T is also ergodic. The proof is complete.

3.1.2 Random Dynamical Systems

Let $\mathbf{X}$ be an arbitrary Banach space. We now introduce a notion of measurable random dynamical systems on $\mathbf{X}$, see [6].

Definition 3.3 A *measurable random dynamical system* on $(\mathbf{X}, \mathcal{B}(\mathbf{X}))$ over a metric dynamical system $(\Omega, \mathcal{F}, \mathbb{P}, (\theta_t)_{t\in\mathbb{T}})$ with time $\mathbb{T}$ is a mapping

$$\varphi : \mathbb{T} \times \Omega \times \mathbf{X} \to \mathbf{X}, \quad (t, \omega, x) \mapsto \varphi(t, \omega, x)$$

with the following properties:

(i) *Measurability*: φ is $\mathcal{B}(\mathbb{T}) \otimes \mathcal{F} \otimes \mathcal{B}(\mathbf{X}), \mathcal{B}(\mathbf{X})$-measurable,
(ii) *Cocycle property*: The mappings $\varphi(t, \omega) := \varphi(t, \omega, \cdot) : \mathbf{X} \to \mathbf{X}$ form a *cocycle* over $\theta(\cdot)$, i.e., they satisfy

$$\varphi(0, \omega) = \mathrm{id}_X \quad \text{for all} \quad \omega \in \Omega,$$

$$\varphi(t + s, \omega) = \varphi(t, \theta_s\omega) \circ \varphi(s, \omega) \quad \text{for all} \quad s, t \in \mathbb{T}, \omega \in \Omega.$$

A random dynamical system (θ, φ) is called *linear* if for any $(t, \omega) \in \mathbb{T} \times \Omega$, the map $\Phi(t, \omega) \equiv \varphi(t, \omega) \in \mathcal{L}(\mathbf{X})$, where $\mathcal{L}(\mathbf{X})$ denotes the space of all bounded linear operators from $\mathbf{X}$ into itself.

3.2 Lyapunov Spectrum and Sacker-Sell Spectrum of Linear Random Dynamical Systems

3.2.1 Multiplicative Ergodic Theorem for Finite-Dimensional Linear Random Dynamical Systems

For a linear autonomous difference equation

$$x_{n+1} = Ax_n, \qquad \text{where } A \in \mathbb{R}^{d\times d},$$

the spectrum of A consisting of eigenvalues of A describes completely its dynamic. A generalization of this dynamical description for linear random dynamical systems driven by an ergodic metric dynamical systems $(\Omega, \mathcal{F}, \mathbb{P}, \theta)$

$$x_{n+1} = A(\theta^n \omega) x_n, \qquad \text{where } A(\omega) \in \mathbb{R}^{d \times d}$$

was established by Oseledets in [96]. Nowadays, this celebrated result is called *Multiplicative ergodic theorem.*

Theorem 3.2 (Multiplicative Ergodic Theorem in Finite Dimensional Spaces) *Let* $\Phi : \mathbb{T} \times \Omega \to \mathbb{R}^{d \times d}$ *be a linear random dynamical system on* $\mathbb{R}^d$*. Suppose that* Φ *satisfies the following* integrability condition

$$\sup_{t \in \mathbb{T} \cap [0,1]} \ln^+ \|\Phi(t, \cdot)^{-1}\| \in \mathcal{L}^1(\Omega, \mathcal{F}, \mathbb{P}).$$

Then, there exist a set $\widehat{\Omega}$ *of full probability being invariant under* $(\theta_t)_{t \in \mathbb{T}}$*, finite number of Lyapunov exponents* $\lambda_p < \cdots < \lambda_1$*, and an invariant splitting*

$$\mathbb{R}^d = O_1(\omega) \oplus O_2(\omega) \oplus \cdots \oplus O_p(\omega) \quad \textit{for } \omega \in \widehat{\Omega}$$

such that for $i = 1, \ldots, p$

$$\lim_{t \to \pm\infty} \frac{1}{t} \ln \|\Phi(t, \omega) v\| = \lambda_i \quad \textit{iff} \quad v \in O_i(\omega) \setminus \{0\}.$$

Proof See, e.g., [6, Theorem 3.4.11].

At the end of this subsection, we provide a result of multiplicative erogdic theorem for induced linear random dynamical systems on exterior powers: For $1 \leq k \leq d$, let $\Lambda^k \mathbb{R}^d$, the *k-fold exterior power* of $\mathbb{R}^d$, be the vector space of alternating k-linear forms on the dual space $\mathbb{R}^d$. The space $\Lambda^k \mathbb{R}^d$ can be identified with the set of formal expressions $\sum_{i=1}^m c_i (u_1^{(i)} \wedge \cdots \wedge u_k^{(i)})$ with $m \in \mathbb{N}$, $c_i \in \mathbb{R}$ and $u_j^{(i)} \in \mathbb{R}^d$ if we do computations with the following conventions:

1. *Addition*:

$$u_1 \wedge \cdots \wedge (u_j + \widehat{u}_j) \wedge \cdots \wedge u_k =$$
$$u_1 \wedge \cdots \wedge u_j \wedge \cdots \wedge u_k + u_1 \wedge \cdots \wedge \widehat{u}_j \wedge \cdots \wedge u_k,$$

2. *Scalar multiplication*: $u_1 \wedge \cdots \wedge c u_j \wedge \cdots \wedge u_k = c u_1 \wedge \cdots \wedge u_j \wedge \cdots \wedge u_k$,
3. for any permutation π of $\{1, \ldots, k\}$

$$u_{\pi(1)} \wedge \cdots \wedge u_{\pi(k)} = \text{sign}(\pi) u_1 \wedge \cdots \wedge u_k.$$

The canonical inner product on $\mathbb{R}^d$ induces an inner product $\langle \cdot, \cdot \rangle$ on $\Lambda^k \mathbb{R}^d$ via

$$\langle u_1 \wedge \cdots \wedge u_k, v_1 \wedge \cdots \wedge v_k \rangle := \det(\langle u_i, v_j \rangle)_{k \times k}.$$

Let $\{e_i\}_{i=1\dots,d}$ denote the standard Euclidean orthogonal basic of $\mathbb{R}^d$. For each $k \in \{1, \dots, d\}$, we define

$$e_{i_1 \dots i_k} := e_{i_1} \wedge \cdots \wedge e_{i_k} \qquad \text{for } 1 \le i_1 < \cdots < i_k \le d.$$

Then,

$$\{e_{i_1 \dots i_k} : 1 \le i_1 < \cdots < i_k \le d\}$$

is an orthogonal basis of $\Lambda^k \mathbb{R}^d$.

A random map $A : \Omega \to \mathbb{R}^{d \times d}$ generates a random map $\Lambda^k A : X \to \mathcal{L}(\Lambda^k \mathbb{R}^d)$ defined by

$$\Lambda^k A(\omega)(u_1 \wedge \cdots \wedge u_k) := A(\omega)u_1 \wedge \cdots \wedge A(\omega)u_k,$$

where $\mathcal{L}(\Lambda^k \mathbb{R}^d)$ denotes the space of linear operators from $\Lambda^k \mathbb{R}^d$ into itself. The generated linear random dynamical system by $\Lambda^k A$ on $\Lambda^k \mathbb{R}^d$, which is denoted by $\Lambda^k \Phi_A$, is called *the k fold exterior of* Φ_A.

Theorem 3.3 (Lyapunov Exponents of Induced Linear Random Dynamical Systems on the Exterior Power) *Let* $\gamma_d(A) \le \gamma_{d-1}(A) \le \cdots \le \gamma_1(A)$ *denote the Lyapunov exponents (counting with multiplicity) of the linear random dynamical system* Φ_A *generated by A. Then, the Lyapunov exponents (counting with multiplicity) of* $\Lambda^k \Phi_A$ *are all sums of k of the exponents of* Φ_A*:*

$$\left\{\gamma_{i_1}(A) + \cdots + \gamma_{i_k}(A) : 1 \le i_1 < \cdots < i_k \le d\right\}.$$

Proof See [49, Remark 3.4].

3.2.2 *Exponential Dichotomies and Dichotomy Spectrum for Random Dynamical Systems*

As is presented in the preceding Sect. 3.2.1, muliplicative ergodic theorem managed to provide a spectral theory of random dynamical systems in term of characterizing the average growth rate of cocycles when the time tends to infinity. Our aim in this section is to establish a new spectral theory based on uniformly estimating the growth rate of cocycles. The main core of this theory is the extension of the notion exponential dichotomy for linear nonautonomous differential equations; see Sect. 1.1, to random dynamical systems.

We define invariant projectors and exponential dichotomies as tools to describe hyperbolicity and (un)-stable manifolds of linear random dynamical systems.

Given a linear random dynamical system $((\theta_t)_{t\in\mathbb{T}}, \Phi)$ on $\mathbb{R}^d$, an random linear subspace $M : \Omega \to \mathbb{R}^d$ is called *invariant* if

$$\Phi(t, \omega)M(\omega) = M(\theta_t\omega) \quad \text{for } \omega \in \Omega.$$

Given invariant linear random subspaces M_1, M_2,

$$\omega \mapsto M_1(\omega) \cap M_2(\omega) \quad \text{and} \quad \omega \mapsto M_1(\omega) + M_2(\omega)$$

are also invariant, denoted by $M_1 \cap M_2$ and $M_1 + M_2$, respectively. A finite sum $M_1 + \cdots + M_n$ of invariant linear random subspaces is called *Whitney sum* $M_1 \oplus \cdots \oplus M_n$ if $M_1(\omega) \oplus \cdots \oplus M_n(\omega) = \mathbb{R}^d$ holds for almost all $\omega \in \Omega$.

An *invariant projector* of (θ, φ) is a measurable function $P : \Omega \to \mathbb{R}^{d\times d}$ with

$$P(\omega) = P(\omega)^2 \quad \text{and} \quad P(\theta_t\omega)\Phi(t, \omega) = \Phi(t, \omega)P(\omega) \quad \text{for all } t \in \mathbb{R} \text{ and } \omega \in \Omega\,.$$

The *range*

$$\mathcal{R}(P) := \big\{(\omega, x) \in \Omega \times \mathbb{R}^d : x \in \mathcal{R}P(\omega)\big\}$$

and the *null space*

$$\mathcal{N}(P) := \big\{(\omega, x) \in \Omega \times \mathbb{R}^d : x \in \mathcal{N}P(\omega)\big\}$$

of an invariant projector P are invariant linear random subspaces of (θ, φ) such that $\mathcal{R}(P) \oplus \mathcal{N}(P) = \Omega \times \mathbb{R}^d$.

The following proposition says that, provided ergodicity, the dimensions of the range and the null space of an invariant projector are almost surely constant.

Proposition 3.1 *Let $P : \Omega \to \mathbb{R}^{d\times d}$ be an invariant projector of an ergodic linear random dynamical system (θ, φ). Then*

(i) the mapping $\omega \mapsto \operatorname{rk} P(\omega)$ is measurable, and

(ii) $\operatorname{rk} P(\omega)$ is almost surely constant.

Proof

(i) We first show that the mapping $A \mapsto \dim A$ on $\mathbb{R}^{d\times d}$ is lower semicontinuous. For this purpose, let $\{A_k\}_{k\in\mathbb{N}}$ be a sequence of matrices in $\mathbb{R}^{d\times d}$ which converges to $A \in \mathbb{R}^{d\times d}$, and define $r := \dim A$. Then there exist nonzero vectors $x_1, \dots, x_r$ such that $Ax_1, \dots, Ax_r$ are linearly independent, which imply that $\det[Ax_1, \dots, Ax_r, x_{r+1}, \dots, x_d] \neq 0$ for some vectors $x_{r+1}, \dots, x_d \in \mathbb{R}^d$. Since $\lim_{k\to\infty} A_k = A$, one gets

$$\lim_{k\to\infty} \det[A_k x_1, \dots, A_k x_r, x_{r+1}, \dots, x_d] = \det[Ax_1, \dots, Ax_r, x_{r+1}, \dots, x_d]\,.$$

Hence, there exists a $k_0 \in \mathbb{N}$ such that vectors $A_k x_1, \ldots, A_k x_r$ are linearly independent for $k \geq k_0$, and thus, $\dim A_k \geq r$ for all $k \geq k_0$. Consequently, the lower semicontinuity of the mapping $A \mapsto \dim A$ is proved. Therefore, the map $\mathbb{R}^{d\times d} \to \mathbb{N}$, $A \mapsto \dim A$ is the limit of a monotonically increasing sequence of continuous functions [121] and thus is measurable. The proof of this part is complete.

(ii) By invariance of P, we get that

$$P(\theta_t\omega) = \Phi(t,\omega)P(\omega)\Phi(t,\omega)^{-1},$$

which implies that $\dim P(\theta_t\omega) = \dim P(\omega)$. This together with ergodicity of θ and measurability of the map $\omega \mapsto \dim P(\omega)$ as shown in (i) gives that $\dim P(\omega)$ is almost constant.

According to Proposition 3.1, the rank of an invariant projector P can be defined via

$$\operatorname{rk} P := \dim \mathcal{R}(P) := \dim \mathcal{R}P(\omega) \quad \text{for almost all } \omega \in \Omega\,,$$

and one sets

$$\dim \mathcal{N}(P) := \dim \mathcal{N}P(\omega) \quad \text{for almost all } \omega \in \Omega.$$

The following notion of an exponential dichotomy describes uniform exponential splitting of linear random dynamical systems (cf. Definition 1.2 for a notion of exponential dichotomy for linear nonautonomous differential equations).

Definition 3.4 (Exponential Dichotomy) Let (θ, Φ) be a linear random dynamical system, and let $\gamma \in \mathbb{R}$ and $P_\gamma : \Omega \to \mathbb{R}^{d\times d}$ be an invariant projector of (θ, φ). Then (θ, φ) is said to admit an *exponential dichotomy* with growth rate $\gamma \in \mathbb{R}$, constants $\alpha > 0$, $K \geq 1$, and projector P_γ if for almost all $\omega \in \Omega$, one has

$$\|\Phi(t,\omega)P_\gamma(\omega)\| \leq Ke^{(\gamma-\alpha)t} \quad \text{for all } t \geq 0\,,$$
$$\|\Phi(t,\omega)(\mathrm{id} - P_\gamma(\omega))\| \leq Ke^{(\gamma+\alpha)t} \quad \text{for all } t \leq 0\,.$$

The following proposition shows that the ranges and null spaces of invariant projectors are given by sums of Oseledets subspaces.

Proposition 3.2 *Let (θ, Φ) be an ergodic linear random dynamical system which satisfies the integrability condition of Oseledets multiplicative ergodic theorem; see Theorem 3.2. Let $\lambda_1 > \cdots > \lambda_p$ and $\mathcal{O}_1(\omega), \ldots, \mathcal{O}_p(\omega)$ denote the Lyapunov exponents and the associated Oseledets subspaces of (θ, Φ), respectively, and suppose that Φ admits an exponential dichotomy with growth rate $\gamma \in \mathbb{R}$ and projector P. Then the following statements hold:*

(i) $\gamma \notin \{\lambda_1, \ldots, \lambda_p\}$.

(ii) *Define* $k := \max\left\{i \in \{0, \dots, p\} : \lambda_i > \gamma\right\}$ *with the convention that* $\lambda_0 = \infty$. *Then for almost all* $\omega \in \Omega$, *one has*

$$\mathcal{NP}(\omega) = \bigoplus_{i=1}^{k} O_i(\omega) \quad \textit{and} \quad \mathcal{RP}(\omega) = \bigoplus_{i=k+1}^{p} O_i(\omega)\,.$$

Proof

(i) Suppose to the contrary that $\gamma = \lambda_k$ for some $k \in \{1, \dots, p\}$. Because of the multiplicative ergodic theorem, see Theorem 3.2; we have

$$\lim_{t\to\infty} \frac{1}{t} \ln \|\Phi(t, \omega)v\| = \lambda_k = \gamma \qquad \text{for all } v \in O_k(\omega) \setminus \{0\}\,. \tag{3.1}$$

On the other hand, for all $v \in \mathcal{RP}_\gamma(\omega)$ we get $\|\Phi(t, \omega)v\| \le Ke^{(\gamma-\alpha)t}\|v\|$ for all $t \ge 0$. Thus,

$$\limsup_{t\to\infty} \frac{1}{t} \ln \|\Phi(t, \omega)v\| \le \gamma - \alpha \quad \text{for all } v \in \mathcal{RP}(\omega)\,,$$

which together with (3.1) implies that $O_k(\omega) \cap \mathcal{RP}(\omega) = \{0\}$. Similarly, using the fact that

$$\lim_{t\to-\infty} \frac{1}{t} \ln \|\Phi(t, \omega)v\| = \lambda_k = \gamma \quad \text{for all } v \in O_k(\omega) \setminus \{0\}$$

and Definition 3.4, we obtain that $O_k(\omega) \cap \mathcal{NP}(\omega) = \{0\}$. Consequently, $O_k(\omega) = \{0\}$, and it leads to a contradiction.

(ii) Let $v \in \mathcal{RP}(\omega) \setminus \{0\}$ be arbitrary. Then, according to Definition 3.4 and the definition of k, we obtain that

$$\lim_{t\to\infty} \frac{1}{t} \ln \|\Phi(t, \omega)v\| \le \gamma - \alpha < \lambda_k. \tag{3.2}$$

Now we write v in the form $v = v_i + v_{i+1} + \cdots + v_p$, where $i \in \{1, \dots, p\}$ with $v_i \ne 0$ and $v_j \in O_j(\omega)$ for all $j = i, \dots, p$. Using the fact that for $j \in \{i, \dots, p\}$ with $v_j \ne 0$

$$\lim_{t\to\infty} \frac{1}{t} \ln \|\Phi(t, \omega)v_j\| = \lambda_j \le \lambda_i,$$

we obtain that

$$\lim_{t\to\infty} \frac{1}{t} \ln \|\Phi(t, \omega)v\| = \lambda_i,$$

which together with (3.2) implies that $i \geq k+1$ and therefore $\mathcal{RP}(\omega) \subset \bigoplus_{i=k+1}^{p} O_i(\omega)$. Similarly, we also get that $\mathcal{NP}(\omega) \subset \bigoplus_{i=1}^{k} O_i(\omega)$. On the other hand,

$$\mathbb{R}^d = \mathcal{NP}(\omega) \oplus \mathcal{RP}(\omega) = \bigoplus_{i=1}^{k} O_i(\omega) \oplus \bigoplus_{i=k+1}^{p} O_i(\omega).$$

Consequently, we have $\mathcal{RP}(\omega) = \bigoplus_{i=k+1}^{p} O_i(\omega)$ and $\mathcal{NP}(\omega) \subset \bigoplus_{i=1}^{k} O_i(\omega)$. The proof is complete.

The monotonicity of the exponential function implies the following basic criteria for the existence of exponential dichotomies.

Lemma 3.1 *Suppose that the linear random system (θ, Φ) admits an exponential dichotomy with growth rate γ and projector P_γ. Then the following statements are fulfilled:*

(i) If $P_\gamma \equiv id$ almost surely, then (θ, Φ) admits an exponential dichotomy with growth rate ζ and invariant projector $P_\zeta \equiv id$ for all $\zeta > \gamma$.
(ii) If $P_\gamma \equiv 0$ almost surely, then (θ, Φ) admits an exponential dichotomy with growth rate ζ and invariant projector $P_\zeta \equiv 0$ for all $\zeta < \gamma$.

Given $\gamma \in \mathbb{R}$, a function $g : \mathbb{R} \to \mathbb{R}^d$ is called γ^+*-exponentially bounded* if $\sup_{t\in\mathbb{R}\cap[0,\infty)} \|g(t)\|e^{-\gamma t} < \infty$. Accordingly, one says that a function $g : \mathbb{R} \to \mathbb{R}^d$ is γ^-*-exponentially bounded* if $\sup_{t\in\mathbb{R}\cap(-\infty,0]} \|g(t)\|e^{-\gamma t} < \infty$.

We define for all $\gamma \in \mathbb{R}$

$$\mathcal{S}^\gamma := \left\{(\omega, x) \in \Omega \times \mathbb{R}^d : \Phi(\cdot, \omega)x \text{ is } \gamma^+\text{-exponentially bounded}\right\},$$

and

$$\mathcal{U}^\gamma := \left\{(\omega, x) \in \Omega \times \mathbb{R}^d : \Phi(\cdot, \omega)x \text{ is } \gamma^-\text{-exponentially bounded}\right\}.$$

It is obvious that $\mathcal{S}^\gamma$ and $\mathcal{U}^\gamma$ are linear invariant random sets of (θ, φ), and given $\gamma \leq \zeta$, the relations $\mathcal{S}^\gamma \subset \mathcal{S}^\zeta$ and $\mathcal{U}^\gamma \supset \mathcal{U}^\zeta$ are fulfilled.

The relationship between the projectors of exponential dichotomies with growth rate γ and the sets $\mathcal{S}^\gamma$ and $\mathcal{U}^\gamma$ will now be discussed.

Proposition 3.3 *If the linear random dynamical system (θ, Φ) admits an exponential dichotomy with growth rate γ and projector P_γ, then $\mathcal{N}(P_\gamma) = \mathcal{U}^\gamma$ and $\mathcal{R}(P_\gamma) = \mathcal{S}^\gamma$ almost surely.*

Proof Suppose that (θ, Φ) admits an exponential dichotomy with growth rate γ, constants α, K and projector P_γ. This means that for almost all $\omega \in \Omega$, one has

$$\|\Phi(t, \omega)P_\gamma(\omega)\| \leq K e^{(\gamma-\alpha)t} \quad \text{for all } t \geq 0,$$

$$\|\Phi(t,\omega)(\mathrm{id}-P_\gamma(\omega))\| \le Ke^{(\gamma+\alpha)t} \quad \text{for all } t \le 0\,.$$

We now prove the relation $\mathcal{N}(P_\gamma) = \mathcal{U}^\gamma$.
($\supseteq$) Choose $(\omega, x) \in \mathcal{U}^\gamma$ arbitrarily. This implies $\|\Phi(t,\omega)x\| \le Ce^{\gamma t}$ for all $t \le 0$ with some real constant $C > 0$. Write $x = x_1 + x_2$ with $x_1 \in \mathcal{R}P_\gamma(\omega)$ and $x_2 \in \mathcal{N}P_\gamma(\omega)$. Hence, for all $t \le 0$,

$$\|x_1\| = \|\Phi(-t,\theta_t\omega)\Phi(t,\omega)P_\gamma(\omega)x\| = \|\Phi(-t,\theta_t\omega)P_\gamma(\theta_t\omega)\Phi(t,\omega)x\|$$

$$\le Ke^{-(\gamma-\alpha)t}\|\Phi(t,\omega)x\| \le CKe^{-(\gamma-\alpha)t}e^{\gamma t} = CKe^{\alpha t}\,.$$

The right-hand side of this inequality converges to zero in the limit $t \to -\infty$. This implies $x_1 = 0$, and thus, $(\omega, x) \in \mathcal{N}(P_\gamma)$.
($\subseteq$) Choose $(\omega, x) \in \mathcal{N}(P_\gamma)$. Thus, for all $t \le 0$, the relation $\|\Phi(t,\omega)x\| \le Ke^{(\gamma+\alpha)t}\|x\|$ is fulfilled. This means that $\Phi(\cdot,\omega)x$ is γ^--exponentially bounded. The proof of statement concerning the range of the projector is treated analogously.

We introduce the dichotomy spectrum for random dynamical systems (cf. Definition 1.3 for the notion of dichotomy spectrum for linear nonautonomous differential equations). For the definition of the dichotomy spectrum, it is crucial for which growth rates, a linear random dynamical system (θ, Φ) admits an exponential dichotomy. The growth rates $\gamma = \pm\infty$ are not excluded from our considerations; in particular, one says that (θ, Φ) admits an exponential dichotomy with growth rate ∞ if there exists a $\gamma \in \mathbb{R}$ such that (θ, Φ) admits an exponential dichotomy with growth rate γ and projector $P_\gamma \equiv \mathrm{id}$. Accordingly, one says that (θ, Φ) admits an exponential dichotomy with growth rate $-\infty$ if there exists a $\gamma \in \mathbb{R}$ such that (θ, Φ) admits an exponential dichotomy with growth rate γ and projector $P_\gamma \equiv 0$.

Definition 3.5 (Dichotomy Spectrum) Consider the linear random dynamical system (θ, Φ). Then the *dichotomy spectrum* (θ, Φ) is defined by

$$\Sigma := \left\{\gamma \in \overline{\mathbb{R}} : (\theta, \Phi) \text{ does not admit an ED with growth rate } \gamma\right\}.$$

The corresponding *resolvent sets* is defined by $\rho := \overline{\mathbb{R}} \setminus \Sigma$.

The aim of the following lemma is to analyze the topological structure of the resolvent sets.

Lemma 3.2 *Consider the resolvent set ρ of a linear random dynamical system (θ, Φ). Then $\rho \cap \mathbb{R}$ is open. More precisely, for all $\gamma \in \rho \cap \mathbb{R}$, there exists an $\varepsilon > 0$ such that $B_\varepsilon(\gamma) \subset \rho$. Furthermore, the relation $\mathrm{rk}\, P_\zeta = \mathrm{rk}\, P_\gamma$ is (almost surely) fulfilled for all $\zeta \in B_\varepsilon(\gamma)$ and every invariant projector P_γ and P_ζ of the exponential dichotomies of (θ, Φ) with growth rates γ and ζ, respectively.*

Proof Choose $\gamma \in \rho$ arbitrarily. Since (θ, Φ) admits an exponential dichotomy with growth rate γ, there exist an invariant projector P_γ and constants $\alpha > 0$, $K \ge 1$ such that for almost all $\omega \in \Omega$, one has

$$\|\Phi(t,\omega)P_\gamma(\omega)\| \le Ke^{(\gamma-\alpha)t} \quad \text{for all } t \ge 0,$$

$$\|\Phi(t,\omega)(\mathrm{id} - P_\gamma(\omega))\| \le Ke^{(\gamma+\alpha)t} \quad \text{for all } t \le 0.$$

Set $\varepsilon := \frac{1}{2}\alpha$, and choose $\zeta \in B_\varepsilon(\gamma)$. It follows that for almost all $\omega \in \Omega$,

$$\|\Phi(t,\omega)P_\gamma(\omega)\| \le Ke^{(\zeta-\frac{\alpha}{2})t} \quad \text{for all } t \ge 0,$$

$$\|\Phi(t,\omega)(\mathrm{id} - P_\gamma(\omega))\| \le Ke^{(\zeta+\frac{\alpha}{2})t} \quad \text{for all } t \le 0.$$

This yields $\zeta \in \rho$, and it follows that $\operatorname{rk} P_\zeta = \operatorname{rk} P_\gamma$ for any projector P_ζ of the exponential dichotomy with growth rate ζ. This finishes the proof of this lemma.

Lemma 3.3 *Consider the resolvent set ρ of a linear random dynamical system (θ, Φ), and let $\gamma_1, \gamma_2 \in \rho \cap \mathbb{R}$ such that $\gamma_1 < \gamma_2$. Moreover, choose invariant projectors P_{γ_1} and P_{γ_2} for the corresponding exponential dichotomies with growth rates γ_1 and γ_2. Then the relation* $\operatorname{rk} P_{\gamma_1} \le \operatorname{rk} P_{\gamma_2}$ *holds. In addition, $[\gamma_1, \gamma_2] \subset \rho$ is fulfilled if and only if* $\operatorname{rk} P_{\gamma_1} = \operatorname{rk} P_{\gamma_2}$.

Proof The relation $\operatorname{rk} P_{\gamma_1} \le \operatorname{rk} P_{\gamma_2}$ is a direct consequence of Proposition 3.3, since $\mathcal{S}^{\gamma_1} \subset \mathcal{S}^{\gamma_2}$ and $\mathcal{U}^{\gamma_1} \supset \mathcal{U}^{\gamma_2}$. Assume now that $[\gamma_1, \gamma_2] \subset \rho$. Arguing contrapositively, suppose that $\operatorname{rk} P_{\gamma_1} \ne \operatorname{rk} P_{\gamma_2}$, and choose invariant projectors P_γ, $\gamma \in (\gamma_1, \gamma_2)$, for the exponential dichotomies of (θ, Φ) with growth rate γ. Define

$$\zeta_0 := \sup \left\{ \zeta \in [\gamma_1, \gamma_2] : \operatorname{rk} P_\zeta \ne \operatorname{rk} P_{\gamma_2} \right\}.$$

Due to Lemma 3.2, there exists an $\varepsilon > 0$ such that $\operatorname{rk} P_{\zeta_0} = \operatorname{rk} P_\zeta$ for all $\zeta \in B_\varepsilon(\zeta_0)$. This is a contradiction to the definition of ζ_0. Conversely, let $\operatorname{rk} P_{\gamma_1} = \operatorname{rk} P_{\gamma_2}$. Because of $\operatorname{rk} P_{\gamma_1} = \operatorname{rk} P_{\gamma_2}$, Proposition 3.3 yields $\mathcal{N}(P_{\gamma_1}) = \mathcal{N}(P_{\gamma_2})$ almost surely, and P_{γ_2} is an invariant projector of the exponential dichotomy with growth rate γ_1. Thus, one obtains for almost all $\omega \in \Omega$,

$$\|\Phi(t,\omega)P_{\gamma_2}(\omega)\| \le K_1 e^{(\gamma_1-\alpha_1)t} \quad \text{for all } t \ge 0$$

for some $K_1 \ge 1$ and $\alpha_1 > 0$. P_{γ_2} is also projector of the exponential dichotomy on $\mathbb{R}_0^-$ with growth rate γ_2. Hence, for almost all $\omega \in \Omega$, one gets

$$\left\|\Phi(t,\omega)(\mathrm{id} - P_{\gamma_2}(\omega))\right\| \le K_2 e^{(\gamma_2+\alpha_2)t} \quad \text{for all } t \le 0$$

with some $K_2 \ge 1$ and $\alpha_2 > 0$. For all $\gamma \in [\gamma_1, \gamma_2]$, these two inequalities imply by setting $K := \max\{K_1, K_2\}$ and $\alpha := \min\{\alpha_1, \alpha_2\}$ that for almost all $\omega \in \Omega$,

$$\|\Phi(t,\omega)P_{\gamma_2}(\omega)\| \le Ke^{(\gamma-\alpha)t} \quad \text{for all } t \ge 0,$$

$$\|\Phi(t,\omega)(\mathrm{id} - P_{\gamma_2}(\omega))\| \le Ke^{(\gamma+\alpha)t} \quad \text{for all } t \le 0.$$

This means that $\gamma \in \rho$, and thus, $[\gamma_1, \gamma_2] \subset \rho$.

For an arbitrarily chosen $a \in \mathbb{R}$, define

$$[-\infty, a] := (-\infty, a] \cup \{-\infty\}, \qquad [a, \infty] := [a, \infty) \cup \{\infty\}$$

and

$$[-\infty, -\infty] := \{-\infty\}, \qquad [\infty, \infty] := \{\infty\}, \qquad [-\infty, \infty] := \overline{\mathbb{R}}.$$

The following *Spectral Theorem* describes that the dichotomy spectrum consists of at least one and at most d closed intervals.

Theorem 3.4 (Spectral Theorem) *Let (θ, Φ) be a linear random dynamical system with dichotomy spectrum Σ. Then there exists an $n \in \{1, \dots, d\}$ such that*

$$\Sigma = [a_1, b_1] \cup \dots \cup [a_n, b_n]$$

with $-\infty \le a_1 \le b_1 < a_2 \le b_2 < \dots < a_n \le b_n \le \infty$.

Proof Due to Lemma 3.2, the resolvent set $\rho \cap \mathbb{R}$ is open. Thus, $\Sigma \cap \mathbb{R}$ is the disjoint union of closed intervals. The relation $(-\infty, b_1] \subset \Sigma$ implies $[-\infty, b_1] \subset \Sigma$, because the assumption of the existence of a $\gamma \in \mathbb{R}$ such that (θ, φ) admits an exponential dichotomy with growth rate γ and projector $P_\gamma \equiv 0$ leads to $(-\infty, \gamma] \subset \rho$ using Lemma 3.1, and this is a contradiction. Analogously, it follows from $[a_n, \infty) \subset \Sigma$ that $[a_n, \infty] \subset \Sigma$. To show the relation $n \le d$, assume to the contrary that $n \ge d + 1$. Thus, there exist

$$\zeta_1 < \zeta_2 < \dots < \zeta_d \in \rho$$

such that the $d + 1$ intervals $(-\infty, \zeta_1)$, (ζ_1, ζ_2), $\dots$, (ζ_d, ∞) have nonempty intersection with the spectrum Σ. It follows from Lemma 3.3 that

$$0 \le \operatorname{rk} P_{\zeta_1} < \operatorname{rk} P_{\zeta_2} < \dots < \operatorname{rk} P_{\zeta_d} \le d$$

is fulfilled for invariant projectors P_{ζ_i} of the exponential dichotomy with growth rate ζ_i, $i \in \{1, \dots, n\}$. This implies either $\operatorname{rk} P_{\zeta_1} = 0$ or $\operatorname{rk} P_{\zeta_d} = d$. Thus, either

$$[-\infty, \zeta_1] \cap \Sigma = \emptyset \quad \text{or} \quad [\zeta_d, \infty] \cap \Sigma = \emptyset$$

is fulfilled, and this is a contradiction. To show $n \ge 1$, assume that $\Sigma = \emptyset$. This implies $\{-\infty, \infty\} \subset \rho$. Thus, there exist $\zeta_1, \zeta_2 \in \mathbb{R}$ such that (θ, φ) admits an exponential dichotomy with growth rate ζ_1 and projector $P_{\zeta_1} \equiv 0$ and an exponential dichotomy with growth rate ζ_2 and projector $P_{\zeta_2} \equiv \mathrm{id}$. Applying Lemma 3.3, one gets $(\zeta_1, \zeta_2) \cap \Sigma \ne \emptyset$. This contradiction yields $n \ge 1$ and finishes the proof of the theorem.

Each spectral interval is associated with a so-called spectral manifold, which generalizes the stable and unstable manifolds obtained by the ranges and null spaces of invariant projectors of exponential dichotomies.

Theorem 3.5 (Spectral Manifolds) *Consider the dichotomy spectrum*

$$\Sigma = [a_1, b_1] \cup \cdots \cup [a_n, b_n]$$

of the linear random dynamical system (θ, Φ), *and define the invariant projectors* $P_{\gamma_0} := 0$, $P_{\gamma_n} :=$ *id, and for* $i \in \{1, \ldots, n-1\}$, *choose* $\gamma_i \in (b_i, a_{i+1})$ *and projectors* P_{γ_i} *of the nonhyperbolic exponential dichotomy of* (θ, φ) *with growth rate* γ_i. *Then the sets*

$$\mathcal{W}_i := \mathcal{R}(P_{\gamma_i}) \cap \mathcal{N}(P_{\gamma_{i-1}}) \quad \text{for all } i \in \{1, \ldots, n\}$$

are fiber-wise linear subset of $\mathbb{R}^d$, *the so-called* spectral manifolds, *such that*

$$\mathcal{W}_1 \oplus \cdots \oplus \mathcal{W}_n = \Omega \times \mathbb{R}^d$$

and $\mathcal{W}_i \neq \Omega \times \{0\}$ *for* $i \in \{1, \ldots, n\}$.

Proof The sets $\mathcal{W}_1, \ldots, \mathcal{W}_n$ obviously have linear fibers. Suppose that there exists an $i \in \{1, \ldots, n\}$ with $\mathcal{W}_i = \mathbb{R} \times \{0\}$. In case $i = 1$ or $i = n$, Lemma 3.1 implies $[-\infty, \gamma_1] \cap \Sigma = \emptyset$ or $[\gamma_{n-1}, \infty] \cap \Sigma = \emptyset$, and this is a contradiction. In case $1 < i < n$, due to Lemma 3.3, one obtains

$$\begin{aligned} \dim \mathcal{W}_i &= \dim\big(\mathcal{R}(P_{\gamma_i}) \cap \mathcal{N}(P_{\gamma_{i-1}})\big) \\ &= \operatorname{rk} P_{\gamma_i} + d - \operatorname{rk} P_{\gamma_{i-1}} - \dim\big(\mathcal{R}(P_{\gamma_i}) + \mathcal{N}(P_{\gamma_{i-1}})\big) \\ &\geq 1 , \end{aligned}$$

and this is also a contradiction. Now the relation $\mathcal{W}_1 \oplus \cdots \oplus \mathcal{W}_n = \Omega \times \mathbb{R}^d$ will be proved. For $1 \leq i < j \leq n$, due to Proposition 3.3, the relations $\mathcal{W}_i \subset \mathcal{R}(P_{\gamma_i})$ and $\mathcal{W}_j \subset \mathcal{N}(P_{\gamma_{j-1}}) \subset \mathcal{N}(P_{\gamma_i})$ are fulfilled. This yields

$$\mathcal{W}_i \cap \mathcal{W}_j \subset \mathcal{R}(P_{\gamma_i}) \cap \mathcal{N}(P_{\gamma_i}) = \mathbb{R} \times \{0\} ,$$

and one obtains

$$\begin{aligned} \Omega \times \mathbb{R}^d &= \mathcal{W}_1 + \mathcal{N}(P_{\gamma_1}) = \mathcal{W}_1 + \mathcal{N}(P_{\gamma_1}) \cap \big(\mathcal{R}(P_{\gamma_2}) + \mathcal{N}(P_{\gamma_2})\big) \\ &= \mathcal{W}_1 + \mathcal{N}(P_{\gamma_1}) \cap \mathcal{R}(P_{\gamma_2}) + \mathcal{N}(P_{\gamma_2}) = \mathcal{W}_1 + \mathcal{W}_2 + \mathcal{N}(P_{\gamma_2}) . \end{aligned}$$

Here, the fact that linear subspaces $E, F, G \subset \mathbb{R}^d$ with $E \supset G$ fulfill $E \cap (F + G) = (E \cap F) + G$ was used. It follows inductively that

$$\Omega \times \mathbb{R}^d = \mathcal{W}_1 + \cdots + \mathcal{W}_n + \mathcal{N}(P_{\gamma_n}) = \mathcal{W}_1 + \cdots + \mathcal{W}_n \,.$$

This finishes the proof of this theorem.

Remark 3.1 If the linear random dynamical system (θ, Φ) under consideration fulfills the conditions of the multiplicative ergodic theorem, then Proposition 3.1 implies that the spectral manifolds $\mathcal{W}_i$ of the above theorem are given by Whitney sums of Oseledets subspaces.

The remaining part of this section on the dichotomy spectrum will be devoted to studying boundedness properties of the spectrum. Firstly, a criterion for boundedness from above and below is provided by the following proposition.

Proposition 3.4 *Consider a linear random dynamical system (θ, Φ), let Σ denote the dichotomy spectrum of (θ, Φ), and define*

$$\alpha^{\pm}(\omega) := \begin{cases} \ln^+\left(\|\Phi(1, \omega)^{\pm 1}\|\right) & : \mathbb{T} = \mathbb{Z}\,, \\ \ln^+\left(\sup_{t \in [0,1]} \|\Phi(t, \omega)^{\pm 1}\|\right) & : \mathbb{T} = \mathbb{R}\,. \end{cases}$$

Then Σ is bounded from above if and only if

$$\operatorname*{ess\,sup}_{\omega \in \Omega} \alpha^+(\omega) < \infty\,,$$

and Σ is bounded from below if and only if

$$\operatorname*{ess\,sup}_{\omega \in \Omega} \alpha^-(\omega) < \infty\,.$$

Consequently, if the dichotomy spectrum Σ is bounded, then Φ satisfies the integrability condition of the multiplicative ergodic theorem.

Proof Suppose that Σ is bounded from above. Then there exist $K > 0$ and $\alpha \in \mathbb{R}$ such that

$$\|\Phi(t, \omega)\| \le K e^{\alpha t} \quad \text{for almost all } \omega \in \Omega\,,$$

which implies that $\operatorname{ess\,sup}_{\omega \in \Omega} \alpha^+(\omega) \le K e^{|\alpha|}$. On the other hand, suppose that $\operatorname{ess\,sup}_{\omega \in \Omega} \alpha(\omega) < \infty$. Then there exists a measurable set U of probability 1 such that for all $\omega \in \Omega$ we have $\alpha^+(\omega) \le e^{\alpha}$ for some positive number α. Define

$$\widetilde{\Omega} := \bigcap_{n \in \mathbb{Z}} \theta^n U\,.$$

Due to the measure preserving property of θ, we get that $\mathbb{P}(\widetilde{\Omega}) = 1$. Let $\gamma > \alpha$ be arbitrary. Then for all $\omega \in \widetilde{\Omega}$, we have

$$\|\Phi(t,\omega)\| \le e^{\alpha t+\alpha} \quad \text{for all } t > 0\,,$$

which implies that $\gamma \in \mathbb{R} \setminus \Sigma$. Hence, $\Sigma \subset (-\infty, \alpha]$. Similarly, we get that Σ is bounded from below if and only if $\operatorname{ess\,sup}_{\omega\in\Omega} \alpha^-(\omega) < \infty$. This finishes the proof of this proposition.

The following example shows that there exist linear random dynamical systems which satisfy the integrability condition of the multiplicative ergodic theorem, but which have no bounded dichotomy spectrum.

Example 3.2 Let $(\Omega, \mathcal{F}, \mathbb{P})$ be a probability space and $\theta : \mathbb{R} \times \Omega \to \Omega$ be a metric dynamical system which is ergodic and nonatomic. Then there exists, by using [67, Lemma 2, p. 71], a measurable set U of the form

$$U = \bigcup_{k=1}^{\infty} \bigcup_{j=0}^{k} \theta_j U_k\,,$$

where U_i, $i \in \mathbb{N}$, are measurable sets such that

(i) for all $k, \ell \in \mathbb{N}$, $i \in \{0, \dots, k\}$ and $j \in \{0, \dots, \ell\}$, we have

$$\theta_j U_k \cap \theta_i U_\ell = \emptyset \quad \text{whenever } k \neq \ell \text{ or } i \neq j\,,$$

(ii) $0 < \mathbb{P}(U_k) \le \frac{1}{k^3}$ for all $k \in \mathbb{N}$.

We now define a random variable $a : \Omega \to \mathbb{R}$ by

$$a(\omega) := \begin{cases} 1 : \omega \in \Omega \setminus U\,, \\ k : k \text{ is even and } \omega \in \theta_j U_k\,, \\ \frac{1}{k} : k \text{ is odd and } \omega \in \theta_j U_k\,. \end{cases}$$

Using the random variable a, we define a discrete-time scalar linear random dynamical system $\Phi : \mathbb{Z} \times \Omega \to \mathbb{R}$ by

$$\Phi(t,\omega) = \begin{cases} a(\theta_{t-1}\omega) \cdots a(\omega) & : t \ge 1\,, \\ 1 & : t = 0\,, \\ a(\theta_{-1}\omega)^{-1} \cdots a(\theta_t\omega)^{-1} & : t \le -1\,. \end{cases}$$

A direct computation yields that

$$\mathbb{E} \ln^+(\|\Phi(1,\omega)\|) = \sum_{k=1}^{\infty} (2k+1)\mathbb{P}(U_{2k}) \ln(2k) \le \sum_{k=1}^{\infty} (2k+1) \frac{\ln(2k)}{8k^3} < \infty\,,$$

and

$$\mathbb{E}\ln^+(\|\Phi(1,\omega)^{-1}\|) = \sum_{k=0}^{\infty}(2k+2)\mathbb{P}(U_{2k+1})\ln(2k+1)$$
$$\leq \sum_{k=1}^{\infty}(2k+2)\frac{\ln(2k+1)}{(2k+1)^3} < \infty\,.$$

Then the linear system Φ satisfies the integrability condition of the multiplicative ergodic theorem. The fact that the dichotomy spectrum of Φ is unbounded from above follows from

$$\|\Phi(n,\omega)\| \geq k^n \quad \text{for all } \omega \in U_k \text{ with } k \text{ even and } 0 \leq n \leq k\,.$$

Similarly, one can prove that the spectrum is unbounded from below.

3.3 Explicit Lyapunov Exponent of Planar Bilinear Stochastic Differential Planar Equations

Taking linear stochastic differential equations as important examples of linear random dynamical systems, our aim in this section is to establish an explicit formula for the top Lyapunov exponent of planar linear stochastic differential equations of the form

$$\begin{pmatrix} dx_t \\ dy_t \end{pmatrix} = A\begin{pmatrix} x_t \\ y_t \end{pmatrix} dt + B\begin{pmatrix} x_t \\ y_t \end{pmatrix} \circ dW_t, \tag{3.3}$$

where $A, B \in \mathbb{R}^{2\times 2}$. The key ingredients are the Furstenberg-Khasminskii formula for Lyapunov exponent (Sect. 3.3.1) and the explicit formulas for the stationary distribution of the associated flow on the projective space (Sect. 3.3.2). For a clearer presentation, we consider two separated cases: the coefficients A and B of (3.3) do not (case (i)) or do (case (ii)) satisfy the Hörmander hypoellipticity condition.

Concerning the case (i), we are able to write explicitly the solutions of (3.3) and to use the strong law of large numbers for martingales to compute explicitly the top Lyapunov exponent of (3.3). Meanwhile, for the case (ii), we first use the Furstenberg-Khasminskii formula to represent the top Lyapunov exponent as an integral of a function involving coefficients A and B over the stationary distribution of the induced flow of (3.3) on the unit circle. Finally, we compute explicitly the stationary distribution by solving the Fokker-Planck equation associated with the induced flow on the unit circle. To do this, depending on the Jordan normal form of the diffusion coefficient B, we can partition the unit circle on some open intervals such that on each open interval, the associated Fokker-Planck equation is a solvable ordinary differential equation (note that in general Fokker-Planck equations are implicit ordinary differential equations). This procedure is also used in [74, 75] to establish explicit formulas for the top Lyapunov exponent and the rotation number of (3.3) in the case that the Jordan normal form of B is $\begin{pmatrix} \alpha & 0 \\ 1 & \alpha \end{pmatrix}$, where $\alpha \in \mathbb{R}$.

3.3.1 *Furtensberg-Khamsminskii Formula for Lyapunov Exponent of Bilinear Stochastic Differential Equations*

Consider a planar linear stochastic differential equation of the form

$$\begin{pmatrix} dx_t \\ dy_t \end{pmatrix} = A \begin{pmatrix} x_t \\ y_t \end{pmatrix} dt + B \begin{pmatrix} x_t \\ y_t \end{pmatrix} \circ dW_t, \tag{3.4}$$

where $\begin{pmatrix} x_t \\ y_t \end{pmatrix} \in \mathbb{R}^2$ and $A, B \in \mathbb{R}^{2\times 2}$. Let $\Phi_{A,B}(t, \xi)$ denote the solution of (3.4) with $(x_0, y_0)^{\mathrm{T}} = \xi \in \mathbb{R}^2 \setminus \{0\}$. Then, the *top sample path Lyapunov exponent* $\lambda_{A,B}$ of (3.4) is defined by

$$\lambda_{A,B} = \lim_{t\to\infty} \frac{1}{t} \log \|\Phi_{A,B}(t, \cdot)\| \qquad \text{a.s.},$$

see, e.g., [6]. To gain a formula to compute $\lambda_{A,B}$, we rewrite equation (3.4) in its polar coordinates by defining $r_t := \sqrt{x_t^2 + y_t^2}$ and $s_t := \left(\frac{x_t}{r_t}, \frac{y_t}{r_t}\right)^{\mathrm{T}}$. Using Ito's formula, see e.g. [82], we obtain

$$dr_t = \overline{f}_A(s_t) r_t \, dt + \overline{f}_B(s_t) r_t \circ dW_t, \quad ds_t = \overline{g}_A(s_t)\, dt + \overline{g}_B(s_t) \circ dW_t,$$

where for a matrix $M = \begin{pmatrix} m_{11} & m_{12} \\ m_{21} & m_{22} \end{pmatrix} \in \mathbb{R}^{2\times 2}$, we define

$$\overline{f}_M(s) := \langle s, Ms\rangle \quad \text{and} \quad \overline{g}_M(s) := Ms - \overline{f}_M(s)s \qquad \text{for } s \in \mathbb{S}^1.$$

By identifying φ_t and $-\varphi_t$, the angular motion is in fact a motion on one-dimensional projective space. Writing $s_t = \begin{pmatrix} \cos\varphi_t \\ \sin\varphi_t \end{pmatrix}$, where $\varphi_t \in \left[-\frac{\pi}{2}, \frac{\pi}{2}\right)$, leads to

$$dr_t = f_A(\varphi_t) r_t \, dt + f_B(\varphi_t) r_t \circ dW_t, \quad d\varphi_t = g_A(\varphi_t)\, dt + g_B(\varphi_t) \circ dW_t,$$

where for a matrix $M = \begin{pmatrix} m_{11} & m_{12} \\ m_{21} & m_{22} \end{pmatrix} \in \mathbb{R}^{2\times 2}$

$$f_M(\varphi) := m_{11}\cos^2\varphi + m_{22}\sin^2\varphi + (m_{12}+m_{21})\cos\varphi\sin\varphi,$$

$$g_M(\varphi) := (m_{22}-m_{11})\cos\varphi\sin\varphi + m_{21}\cos^2\varphi - m_{12}\sin^2\varphi.$$

Now we recall the well-known *Furstenberg-Khasminskii formula* for the top Lyapunov exponent of (3.4).

Theorem 3.6 (Furstenberg-Khasminskii Formula) *Suppose that the following nondegeneracy condition holds:*

(H) There is no $s \in \mathbb{S}^1$ such that

$$As = \langle As, s\rangle s \quad \text{and} \quad Bs = \langle Bs, s\rangle s.$$

Then, the top Lyapunov exponent $\lambda_{A,B}$ of (3.4) is given by

$$\lambda_{A,B} = \int_{-\frac{\pi}{2}}^{\frac{\pi}{2}} \left(f_A(\varphi) + \frac{1}{2} f_B'(\varphi) g_B(\varphi) \right) p(\varphi)\, d\varphi,$$

where $p : [-\frac{\pi}{2}, \frac{\pi}{2}] \to \mathbb{R}_{\geq 0}$ is a smooth density function satisfying that $p(-\frac{\pi}{2}) = p(\frac{\pi}{2})$, $\int_{-\frac{\pi}{2}}^{\frac{\pi}{2}} p(\varphi)\, d\varphi = 1$ and the following differential equation

$$\frac{1}{2} g_B(\varphi)^2 p'(\varphi) = \left(g_A(\varphi) - \frac{1}{2} g_B'(\varphi) g_B(\varphi) \right) p(\varphi) + C, \tag{3.5}$$

where C is a constant.

Proof See, e.g., [75, page 34–37].

3.3.2 Explicit Formulas for Top Lyapunov Exponents

Our aim in this section is to establish explicit formulas for the top Lyapunov exponents of planar linear stochastic differential equations. For this purpose, we divide this section into two subsections. In Sect. 3.3.3, we consider degenerated linear stochastic differential equations, i.e., equations in which the drift and the diffusion coefficients do not fulfill condition (H) of Theorem 3.6. For systems satisfying condition (H), we give explicit formulas for their top Lyapunov exponents in Sect. 3.3.4.

3.3.3 Degenerated Linear Stochastic Differential Equations

Suppose that $A, B \in \mathbb{R}^{2\times 2}$ do not satisfy condition (H). Therefore, there exists $s \in \mathbb{S}^1$ such that

$$As = \langle As, s\rangle s \quad \text{and} \quad Bs = \langle Bs, s\rangle s,$$

which implies that $s \in \mathbb{S}^1$ is a common real eigenvector of A and B. Thus, for the orthogonal matrix $T \in \mathbb{R}^{2\times 2}$ defined by $Te_1 = s$ and $Te_2 := s^\perp$, we have $T^{-1}AT$ and $T^{-1}BT$ are lower triangular matrices. Hence, any degenerated linear stochastic differential equation can be transformed to a lower triangular linear stochastic differential equation, and our aim in this subsection is to give an explicit formula for Lyapunov exponent of this class of linear stochastic differential equations. Before going to the main result in this subsection, we need the following preparatory lemma.

Lemma 3.4 *Let $\alpha, \beta \in \mathbb{R}$ be arbitrary and $(W_t)_{t\in\mathbb{R}}$ be a Brownian motion defined in a probability space $(\Omega, \mathcal{F}, \mathbb{P})$. Then, the following statements hold almost surely:*

(i) $\limsup_{t\to\infty} \frac{1}{t} \log \left| \int_0^t \exp(\alpha s) W_s(\omega)\, ds \right| \leq \alpha.$

(ii) $\lim_{t\to\infty} \frac{1}{t} \log \int_0^t \exp(\alpha s + \beta W_s(\omega))\, ds = \alpha.$

Proof Let ε be an arbitrary positive number. Using strong law of large numbers for martingales, there exists a measurable set $\widehat{\Omega}$ with $\mathbb{P}(\widehat{\Omega}) = 1$ and $\lim_{t\to\infty} \frac{W_t(\omega)}{t} = 0$ for all $\omega \in \widehat{\Omega}$; see, e.g., [23, Appendix A]. Choose and fix an arbitrary $\omega \in \widehat{\Omega}$. Thus, for $\omega \in \widehat{\Omega}$ there exists $T(\varepsilon, \omega) > 0$ such that

$$-\varepsilon s \leq W_s(\omega) \leq \varepsilon s \qquad \text{for all } s \geq T(\varepsilon, \omega).$$

Hence, for all $t \geq T(\varepsilon, \omega)$, we have

$$\int_{T(\varepsilon,\omega)}^t \exp(\alpha s) W_s(\omega)\, ds \leq \varepsilon \int_0^t \exp(\alpha s) s\, ds,$$

which implies that

$$\limsup_{t\to\infty} \frac{1}{t} \log \left| \int_{T(\varepsilon,\omega)}^t \exp(\alpha s) W_s(\omega)\, ds \right| \leq \limsup_{t\to\infty} \frac{1}{t} \log \left(\varepsilon \int_0^t \exp(\alpha s) s\, ds \right) \leq \alpha.$$

Hence, (i) is proved. For all $t \geq T(\varepsilon, \omega)$, we also have

$$\int_{T(\varepsilon,\omega)}^{t} \exp\left(\alpha - |\beta|\varepsilon\right)s\, ds \le \int_{T(\varepsilon,\omega)}^{t} \exp(\alpha s + \beta W_s(\omega))\, ds$$

$$\le \int_{T(\varepsilon,\omega)}^{t} \exp\left(\alpha + |\beta|\varepsilon\right)s\, ds,$$

which implies that

$$\alpha - |\beta|\varepsilon \le \liminf_{t\to\infty} \frac{1}{t} \log \int_0^t \exp(\alpha s + \beta W_s(\omega))\, ds$$

and

$$\alpha + |\beta|\varepsilon \ge \limsup_{t\to\infty} \frac{1}{t} \log \int_0^t \exp(\alpha s + \beta W_s(\omega))\, ds.$$

Letting $\varepsilon \to 0$ yields (ii), and the proof is complete.

Theorem 3.7 (Explicit Formula for the Top Lyapunov Exponents of Degenerated Linear Stochastic Differential Equations) *Consider system* (3.4) *where* A, B *are of the following form*

$$A = \begin{pmatrix} a_{11} & 0 \\ a_{21} & a_{22} \end{pmatrix} \quad \textit{and} \quad B = \begin{pmatrix} b_{11} & 0 \\ b_{21} & b_{22} \end{pmatrix}.$$

Then,

$$\lambda_{A,B} = \rho(A) = \max\left\{a_{11}, a_{22}\right\}.$$

Proof The stochastic differential equation for the first component x_t is

$$dx_t = a_{11}x_t\, dt + b_{11}x_t \circ dW_t,$$

which implies that

$$x_t = \exp\left(a_{11}t + b_{11}W_t\right)x_0.$$

The equation for the second component y_t is

$$dy_t = \left(a_{22} + \frac{b_{22}^2}{2}\right) y_t\, dt + \left(a_{21} + b_{21}\frac{b_{11} + b_{22}}{2}\right) x_t\, dt + (b_{21}x_t + b_{22}y_t)\, dW_t.$$

Using the variation of constants formula, see, e.g., [82, pp. 120]; we obtain that

$$y_t = \exp(a_{22}t + b_{22}W_t)\Big(y_0 + \Big(a_{21} + b_{21}\frac{b_{11}-b_{22}}{2}\Big)x_0\int_0^t \exp(\alpha s + \beta W_s)\,ds$$
$$+b_{21}x_0\int_0^t \exp(\alpha s + \beta W_s)\,dW_s\Big),$$

where $\alpha := a_{11} - a_{22}$ and $\beta := b_{11} - b_{22}$. Since $\lim_{t\to\infty}\frac{W_t}{t} = 0$ it follows that

$$\lim_{t\to\infty}\frac{1}{t}\log|x_t| = a_{11} \quad \text{for } x_0 \neq 0$$

and

$$\lim_{t\to\infty}\frac{1}{t}\log|y_t| = a_{22} \quad \text{for } x_0 = 0,\ y_0 \neq 0.$$

Consequently, $\lambda_{A,B} \geq \rho(A)$. To conclude the proof, it is sufficient to show that $\lambda_{A,B} \leq \rho(A)$. Equivalently, we prove that $\lim_{t\to\infty}\frac{1}{t}\log|y_t| \leq \rho(A)$. To do this, we consider the following separated cases:
Case 1: If $\beta = 0$, then by using Ito's formula, we obtain that

$$\int_0^t \exp(\alpha s)\,dW_s = \exp(\alpha t)W_t - \alpha\int_0^t \exp(\alpha s)W_s\,ds.$$

Therefore,

$$y_t = \exp(a_{22}t + b_{22}W_t)\Big(y_0 + \Big(a_{21} + b_{21}\frac{b_{11}-b_{22}}{2}\Big)x_0\int_0^t \exp(\alpha s)\,ds\Big) -$$
$$\alpha b_{21}x_0\exp(a_{22}t + b_{22}W_t)\int_0^t \exp(\alpha s)W_s\,ds +$$
$$b_{21}x_0\exp(a_{11}t + b_{11}W_t)\,W_t,$$

which together with Lemma 3.4(i) implies that $\limsup_{t\to\infty}\frac{1}{t}\log|y_t| \leq \rho(A)$.
Case 2: If $\beta \neq 0$, then by using Ito's formula, we obtain that

$$\int_0^t \exp(\alpha s + \beta W_s)\,dW_s = \frac{1}{\beta}\big(\exp(\alpha t + \beta W_t) - 1\big)$$
$$-\Big(\frac{\alpha}{\beta} + \frac{\beta}{2}\Big)\int_0^t \exp(\alpha s + \beta W_s)\,ds.$$

Thus,

$$y_t = \Big(y_0 - \frac{b_{21}}{\beta}x_0\Big)\exp(a_{22}t + b_{22}W_t) + \frac{b_{21}}{\beta}x_0\exp(a_{11}t + b_{11}W_t)$$

$$+\big(a_{21} - b_{21}\frac{\alpha}{\beta}\big)x_0\exp(a_{22}t + b_{22}W_t)\int_0^t \exp(\alpha s + \beta W_s)\,ds.$$

By virtue of Lemma 3.4 (ii), $\lim_{t\to\infty}\frac{1}{t}\log|y_t| \le \rho(A)$ and the proof is complete.

3.3.4 Nondegenerated Linear Stochastic Differential Equations

According to Theorem 3.6, the top Lyapunov exponent of a nondegenerated linear stochastic differential equation is given explicitly in terms of the stationary distribution of the induced flow on the unit circle. Hence, to obtain an explicit form of the top Lyapunov exponent, our attempt in this subsection is to solve explicitly nonzero solutions of (3.5). Note that the Eq. (3.5) can be singular in the sense that $g_B(\varphi)$ might be equal to zero for some values of φ. So, to solve (3.5), we distinguish the following types of the form of matrix B (up to a transformation generated by a non-singular matrix):

$$\text{Type I}\ \ B = \alpha\,\text{id}, \qquad \text{Type II}\ \ B = \begin{pmatrix}\alpha & 0\\ 0 & \beta\end{pmatrix},$$

$$\text{Type III}\ B = \begin{pmatrix}\alpha & 0\\ 1 & \alpha\end{pmatrix},\ \text{Type IV}\ B = \begin{pmatrix}\alpha & -\beta\\ \beta & \alpha\end{pmatrix},$$

where $\alpha, \beta \in \mathbb{R}$, and additionally, in Type II, we assume that $\alpha \neq \beta$ and in Type IV we assume that $\beta \neq 0$.

3.3.4.1 Type I

Let B be of Type I. Then, any matrix $A \in \mathbb{R}^{2\times 2}$ commutes with B. In the following lemma, we compute the top Lyapunov exponent of (3.4) when A and B commute. Consequently, we obtain a formula for the top Lyapunov exponent of (3.4) when B is of Type I.

Lemma 3.5 *Suppose that A and B are commutative, i.e.,* $AB = BA$. *Then, for any nonzero initial value* $(x_0, y_0)^{\mathrm{T}} \in \mathbb{R}^2 \setminus \{0\}$, *the characteristic Lyapunov exponent of the solution* $\Phi_{A,B}(t, (x_0, y_0)^{\mathrm{T}})$ *of* (3.4) *is*

$$\limsup_{t\to\infty}\frac{1}{t}\log\left\|\Phi_{A,B}(t,(x_0,y_0)^{\mathrm{T}})\right\| = \lim_{t\to\infty}\frac{1}{t}\log\left\|\exp(tA)(x_0,y_0)^{\mathrm{T}}\right\|. \qquad (3.6)$$

In particular, $\lambda_{A,B} = \rho(A)$.[2]

Proof Since the matrices A and B are commute, it follows that the explicit solution of (3.4) is

$$\Phi_{A,B}(t,(x_0,y_0)^{\mathrm{T}}) = \exp(tA + BW_t)(x_0,y_0)^{\mathrm{T}} = \exp(W_tB)\exp(tA)(x_0,y_0)^{\mathrm{T}},$$

which implies that

$$\begin{aligned}\frac{\left\|\exp(tA)(x_0,y_0)^{\mathrm{T}}\right\|}{\|\exp(-W_tB)\|} &\le \left\|\Phi_{A,B}(t)(x_0,y_0)^{\mathrm{T}}\right\| \\ &\le \|\exp(W_tB)\|\left\|\exp(tA)(x_0,y_0)^{\mathrm{T}}\right\|. \end{aligned} \tag{3.7}$$

Since $\lim_{t\to\infty}\frac{W_t}{t} = 0$ it follows that

$$\lim_{t\to\infty}\frac{1}{t}\log\|\exp(W_tB)\| = \lim_{t\to\infty}\frac{1}{t}\log\|\exp(-W_tB)\| = 0,$$

which together with (3.7) proves (3.6). Since

$$\max_{(x_0,y_0)^{\mathrm{T}}\in\mathbb{R}^2\setminus\{0\}}\lim_{t\to\infty}\frac{1}{t}\log\left\|\exp(tA)(x_0,y_0)^{\mathrm{T}}\right\| = \rho(A),$$

it follows that $\lambda_{A,B} = \rho(A)$. The proof is complete.

Theorem 3.8 (Explicit Formula for the Top Lyapunov Exponents of Linear SDE of Type I) *Consider system* (3.4) *with the drift part* $A = (a_{ij}) \in \mathbb{R}^{2\times 2}$ *and with the diffusion part* $B = \begin{pmatrix}\alpha & 0\\ 0 & \alpha\end{pmatrix}$, *where* $\alpha \in \mathbb{R}$. *Then, the top Lyapunov exponent of* (3.4) *is given by* $\lambda_{A,B} = \rho(A)$.

3.3.4.2 Type II

In this part, we consider the case that B is of Type II. Note that for all diagonal singular matrices $F = \begin{pmatrix} f_1 & 0\\ 0 & f_2\end{pmatrix} \in \mathbb{R}^{2\times 2}$ we have $FBF^{-1} = B$. In the following remark, a suitable diagonal matrix F which enables to simplify the form of the drift term A is found:

[2] For a matrix M, let $\sigma(M)$ denote the set of all complex eigenvalues of M, and let $\rho(M) := \{\max \mathrm{Re}\lambda : \lambda \in \sigma(M)\}$.

Remark 3.2 Let $A = (a_{ij}) \in \mathbb{R}^{2\times 2}$. Then, the pair (A, B) satisfy condition (H) of Theorem 3.6 if and only if $a_{12}a_{21} \neq 0$. Depending on the sign of $a_{12}a_{21}$, we can simplify A as follows:

(a) If $a_{12}a_{21} > 0$, then for $F = \begin{pmatrix} \operatorname{sign}(a_{12})\sqrt{\frac{a_{21}}{a_{12}}} & 0 \\ 0 & 1 \end{pmatrix}$, we have

$$\widehat{A} := FAF^{-1} = \begin{pmatrix} a_{11} & \sqrt{a_{12}a_{21}} \\ \sqrt{a_{12}a_{21}} & a_{22} \end{pmatrix}. \tag{3.8}$$

(b) If $a_{12}a_{21} < 0$, then for $F = \begin{pmatrix} \operatorname{sign}(a_{12})\sqrt{-\frac{a_{21}}{a_{12}}} & 0 \\ 0 & 1 \end{pmatrix}$ we have

$$\widehat{A} := FAF^{-1} = \begin{pmatrix} a_{11} & \sqrt{-a_{12}a_{21}} \\ -\sqrt{-a_{12}a_{21}} & a_{22} \end{pmatrix}. \tag{3.9}$$

Theorem 3.9 (Explicit Formula for the Top Lyapunov Exponents of Linear SDE of Type II) *Consider system* (3.4) *with the drift part* $A = (a_{ij}) \in \mathbb{R}^{2\times 2}$ *and with the diffusion part* $B = \begin{pmatrix} \alpha & 0 \\ 0 & \beta \end{pmatrix}$, *where* $\alpha, \beta \in \mathbb{R}$ *with* $\alpha \neq \beta$. *Then, the following statements hold:*

(a) If $a_{12}a_{21} > 0$, *then*

$$\lambda_{A,B} = \frac{a_{11}+a_{22}}{2} + \frac{\int_0^\infty P_{A,B}(u)\exp\left(-\frac{2\sqrt{a_{12}a_{21}}}{(\alpha-\beta)^2}\frac{1+u^2}{u}\right)du}{2\int_0^\infty Q_{A,B}(u)\exp\left(-\frac{2\sqrt{a_{12}a_{21}}}{(\alpha-\beta)^2}\frac{1+u^2}{u}\right)du},$$

where

$$P_{A,B}(u) := u^{\frac{2(a_{22}-a_{11})}{(\alpha-\beta)^2}-2}\Big((a_{11}-a_{22})(1-u^2) + 4\sqrt{a_{12}a_{21}}u +$$
$$+2(\alpha-\beta)^2\frac{u^2}{1+u^2}\Big),$$

$$Q_{A,B}(u) := u^{\frac{2(a_{22}-a_{11})}{(\alpha-\beta)^2}-2}(1+u^2).$$

(b) If $a_{12}a_{21} < 0$, *then*

$$\lambda_{A,B} = \frac{a_{11}+a_{22}}{2}$$

$$+\frac{1}{C}\int_0^{\infty} R_{A,B}(u)e^{\frac{2\sqrt{-a_{12}a_{21}}}{(\alpha-\beta)^2}\frac{1-u^2}{u}}\int_0^u \frac{v^{-\frac{2(a_{22}-a_{11})}{(\alpha-\beta)^2}}}{1+v^2}$$

$$\times e^{-\frac{2\sqrt{-a_{12}a_{21}}}{(\alpha-\beta)^2}\frac{1-v^2}{v}}\,dv\,du$$

$$\times+\frac{1}{C}\int_{-\infty}^{0} R_{A,B}(u)e^{\frac{2\sqrt{-a_{12}a_{21}}}{(\alpha-\beta)^2}\frac{1-u^2}{u}}\int_{-\infty}^u \frac{|v|^{-\frac{2(a_{22}-a_{11})}{(\alpha-\beta)^2}}}{1+v^2}$$

$$e^{-\frac{2\sqrt{-a_{12}a_{21}}}{(\alpha-\beta)^2}\frac{1-v^2}{v}}\,dv\,du,$$

where

$$R_{A,B}(u) := |u|^{\frac{2(a_{22}-a_{11})}{(\alpha-\beta)^2}-2}\left(\frac{a_{11}-a_{22}}{2}(1-u^2)+(\alpha-\beta)^2\frac{u^2}{1+u^2}\right)$$

and

$$C := \int_0^{\infty} S_{A,B}(u)e^{\frac{2\sqrt{-a_{12}a_{21}}}{(\alpha-\beta)^2}\frac{1-u^2}{u}}\int_0^u \frac{v^{-\frac{2(a_{22}-a_{11})}{(\alpha-\beta)^2}}}{1+v^2}e^{-\frac{2\sqrt{-a_{12}a_{21}}}{(\alpha-\beta)^2}\frac{1-v^2}{v}}\,dv\,du$$

$$+\int_{-\infty}^{0} S_{A,B}(u)e^{\frac{2\sqrt{-a_{12}a_{21}}}{(\alpha-\beta)^2}\frac{1-u^2}{u}}\int_{-\infty}^u \frac{|v|^{-\frac{2(a_{22}-a_{11})}{(\alpha-\beta)^2}}}{1+v^2}e^{-\frac{2\sqrt{-a_{12}a_{21}}}{(\alpha-\beta)^2}\frac{1-v^2}{v}}\,dv\,du,$$

here $S_{A,B}(u) := (1+u^2)|u|^{\frac{2(a_{22}-a_{11})}{(\alpha-\beta)^2}-2}$.

Proof Let $\lambda_{\widehat{A},B}$ denote the top Lyapunov exponent of the following linear stochastic differential equation

$$dX_t = \widehat{A}X_t + BX_t \circ dW_t,$$

where $\widehat{A}$ is defined as in Remark 3.2. Obviously, $\lambda_{A,B} = \lambda_{\widehat{A},B}$ and our aim is to compute $\lambda_{\widehat{A},B}$. For this purpose, by definition of B, we have $g_B(\varphi) = \frac{\beta-\alpha}{2}\sin 2\varphi$. Then, the equation for the stationary distribution of the induced flow on $\mathbb{S}^1$ is

$$\sin^2 2\varphi\, p'(\varphi) = \left(\frac{8g_{\widehat{A}}(\varphi)}{(\beta-\alpha)^2} - 2\sin 4\varphi\right)p(\varphi) + C, \tag{3.10}$$

where C is a constant. Using the explicit formula of $g_{\widehat{A}}(\varphi)$, the differential equation corresponding to the linear part of (3.10) is given by

$$\frac{p'(\varphi)}{p(\varphi)} = \frac{4}{(\alpha-\beta)^2}\left(\frac{\widehat{a}_{22}-\widehat{a}_{11}}{\sin 2\varphi} + \frac{(\widehat{a}_{12}+\widehat{a}_{21})\cos 2\varphi}{\sin^2 2\varphi} + \frac{\widehat{a}_{21}-\widehat{a}_{12}}{\sin^2 2\varphi}\right) - 4\cot 2\varphi,$$

where $\varphi \in [-\frac{\pi}{2}, \frac{\pi}{2}] \setminus \{\pm\frac{\pi}{2}, 0\}$. A solution of the preceding equation on $[-\frac{\pi}{2}, \frac{\pi}{2}] \setminus \{\pm\frac{\pi}{2}, 0\}$ is

$$\Pi(\varphi) := \frac{|\tan\varphi|^{\frac{2(\widehat{a}_{22}-\widehat{a}_{11})}{(\alpha-\beta)^2}}}{\sin^2 2\varphi \exp\left(\frac{2(\widehat{a}_{12}+\widehat{a}_{21})}{(\alpha-\beta)^2 \sin 2\varphi} + \frac{2(\widehat{a}_{21}-\widehat{a}_{12})\cot 2\varphi}{(\alpha-\beta)^2}\right)}. \tag{3.11}$$

By variation of constants formula, we have

$$p(\varphi) = \begin{cases} \Pi(\varphi)\left(\frac{p(\frac{\pi}{4})}{\Pi(\frac{\pi}{4})} + C\int_{\frac{\pi}{4}}^{\varphi} \frac{1}{\sin^2 2s\,\Pi(s)}\,ds\right) & \text{for } \varphi \in (0, \frac{\pi}{2}),\\ \Pi(\varphi)\left(\frac{p(-\frac{\pi}{4})}{\Pi(-\frac{\pi}{4})} + C\int_{-\frac{\pi}{4}}^{\varphi} \frac{1}{\sin^2 2s\,\Pi(s)}\,ds\right) & \text{for } \varphi \in (-\frac{\pi}{2}, 0).\end{cases} \tag{3.12}$$

Now, we consider two separated cases:

(a) Let $\widehat{A}$ be of the form (3.8). Then, by (3.11), the function $\Pi(\varphi)$ is given by

$$\Pi(\varphi) = \frac{|\tan\varphi|^{\frac{2(a_{22}-a_{11})}{(\alpha-\beta)^2}}}{\sin^2 2\varphi \exp\left(\frac{4\sqrt{a_{12}a_{21}}}{(\alpha-\beta)^2}\frac{1}{\sin 2\varphi}\right)},$$

which implies that $\lim_{\varphi\to 0^-}\Pi(\varphi) = \lim_{\varphi\to -\frac{\pi}{2}^+}\Pi(\varphi) = \infty$. Therefore, by boundedness of p and (3.12), we have

$$\frac{p(-\frac{\pi}{4})}{\Pi(-\frac{\pi}{4})} + C\int_{-\frac{\pi}{4}}^{0} \frac{1}{\Pi(s)\sin^2 2s}\,ds = \frac{p(-\frac{\pi}{4})}{\Pi(-\frac{\pi}{4})} + C\int_{-\frac{\pi}{4}}^{-\frac{\pi}{2}} \frac{1}{\Pi(s)\sin^2 2s}\,ds = 0.$$

Consequently, $p(-\frac{\pi}{4}) = C = 0$ and using the fact that $\int_{-\frac{\pi}{2}}^{\frac{\pi}{2}} p(\varphi)\,d\varphi = 1$, we obtain that

$$p(\varphi) = \begin{cases} \frac{\Pi(\varphi)}{\int_0^{\frac{\pi}{2}}\Pi(s)\,ds}, & \text{if } \varphi \in [0, \frac{\pi}{2}],\\ 0, & \text{if } \varphi \in [-\frac{\pi}{2}, 0).\end{cases}$$

In light of Theorem 3.6 and from the fact that

$$f_{\widehat{A}}(\varphi) + \frac{1}{2}f_B'(\varphi)g_B(\varphi) = \frac{a_{11}+a_{22}}{2} + \frac{a_{11}-a_{22}}{2}\cos 2\varphi + \sqrt{a_{12}a_{21}}\sin 2\varphi + \frac{(\alpha-\beta)^2}{4}\sin^2 2\varphi,$$

we have

$$\lambda_{A,B} = \frac{a_{11}+a_{22}}{2} + \frac{\int_0^{\frac{\pi}{2}} \left(\frac{a_{11}-a_{22}}{2}\cos 2\varphi + \sqrt{a_{12}a_{21}}\sin 2\varphi + \frac{(\alpha-\beta)^2}{4}\sin^2 2\varphi\right)\Pi(\varphi)\,d\varphi}{\int_0^{\frac{\pi}{2}} \Pi(\varphi)\,d\varphi}.$$

Changing variable $\varphi = \arctan u$ in the preceding integral completes the proof of this part.

(b) Let $\widehat{A}$ be of the form (3.9). Then, by (3.11), the function $\Pi(\varphi)$ is of the following form

$$\Pi(\varphi) = \frac{|\tan\varphi|^{\frac{2(a_{22}-a_{11})}{(\alpha-\beta)^2}}}{\sin^2 2\varphi \exp\left(-\frac{4\sqrt{-a_{12}a_{21}}}{(\alpha-\beta)^2}\cot 2\varphi\right)},$$

which implies that $\lim_{\varphi\to 0^+}\Pi(\varphi) = \lim_{\varphi\to -\frac{\pi}{2}^+}\Pi(\varphi) = \infty$. Consequently, from boundedness of $p(\varphi)$ and (3.12), we have

$$\frac{p(\frac{\pi}{4})}{\Pi(\frac{\pi}{4})} + C\int_{\frac{\pi}{4}}^0 \frac{1}{\sin^2 2s\,\Pi(s)}\,ds = \frac{p(-\frac{\pi}{4})}{\Pi(-\frac{\pi}{4})} + C\int_{-\frac{\pi}{4}}^{-\frac{\pi}{2}} \frac{1}{\sin^2 2s\,\Pi(s)}\,ds = 0.$$

Thus,

$$p(\varphi) = \begin{cases} C\int_0^{\varphi} \frac{\Pi(\varphi)}{\sin^2 2s\,\Pi(s)}\,ds & \text{for } \varphi\in(0,\frac{\pi}{2}),\\ C\int_{-\frac{\pi}{2}}^{\varphi} \frac{\Pi(\varphi)}{\sin^2 2s\,\Pi(s)}\,ds & \text{for } \varphi\in(-\frac{\pi}{2},0). \end{cases}$$

Since $\int_{-\frac{\pi}{2}}^{\frac{\pi}{2}} p(\varphi)\,d\varphi = 1$, it follows that

$$C = \frac{1}{\int_0^{\frac{\pi}{2}}\int_0^{\varphi} \frac{\Pi(\varphi)}{\sin^2 2s\,\Pi(s)}\,ds\,d\varphi + \int_{-\frac{\pi}{2}}^0\int_{-\frac{\pi}{2}}^{\varphi} \frac{\Pi(\varphi)}{\sin^2 2s\,\Pi(s)}\,ds\,d\varphi}.$$

A direct computation yields that

$$f_{\widehat{A}}(\varphi) + \frac{1}{2}f_B'(\varphi)g_B(\varphi) = \frac{a_{11}+a_{22}}{2} + \frac{a_{11}-a_{22}}{2}\cos 2\varphi + \frac{(\alpha-\beta)^2}{4}\sin^2 2\varphi,$$

which together with Theorem 3.6 implies that

$$\lambda_{A,B} = \frac{a_{11}+a_{22}}{2} + \frac{\int_0^{\frac{\pi}{2}}\left(\frac{a_{11}-a_{22}}{2}\cos 2\varphi + \frac{(\alpha-\beta)^2}{4}\sin^2 2\varphi\right)\int_0^{\varphi} \frac{\Pi(\varphi)}{\sin^2 2s\,\Pi(s)}\,ds\,d\varphi}{\int_0^{\frac{\pi}{2}}\int_0^{\varphi} \frac{\Pi(\varphi)}{\sin^2 2s\,\Pi(s)}\,ds\,d\varphi + \int_{-\frac{\pi}{2}}^0\int_{-\frac{\pi}{2}}^{\varphi} \frac{\Pi(\varphi)}{\sin^2 2s\,\Pi(s)}\,ds\,d\varphi}$$

$$+\frac{\int_{-\frac{\pi}{2}}^{0}\left(\frac{a_{11}-a_{22}}{2}\cos 2\varphi+\frac{(\alpha-\beta)^2}{4}\sin^2 2\varphi\right)\int_{-\frac{\pi}{2}}^{\varphi}\frac{\Pi(\varphi)}{\sin^2 2s\,\Pi(s)}\,ds\,d\varphi}{\int_0^{\frac{\pi}{2}}\int_0^{\varphi}\frac{\Pi(\varphi)}{\sin^2 2s\,\Pi(s)}\,ds\,d\varphi+\int_{-\frac{\pi}{2}}^{0}\int_{-\frac{\pi}{2}}^{\varphi}\frac{\Pi(\varphi)}{\sin^2 2s\,\Pi(s)}\,ds\,d\varphi}.$$

Changing variables $s = \arctan v$ and $\varphi = \arctan u$ in the preceding integral completes the proof.

3.3.4.3 Type III

In this subsection, we recall the result in [74, 75] about explicit formula of Lyapunov exponent of linear stochastic differential equations whose diffusion parts are of Type III. Note that in this case, a pair (A, B) do not satisfy the degeneracy condition (H) if the matrix A is not of lower triangular form, i.e., $a_{12} \neq 0$.

Theorem 3.10 (Explicit Formula for the Lyapunov Exponents of Linear SDE of Type III) *Consider system* (3.4) *with the drift part* $A = (a_{ij}) \in \mathbb{R}^{2\times 2}$ *and with the diffusion part* $B = \begin{pmatrix} \alpha & 0 \\ 1 & \alpha \end{pmatrix}$, *where* $\alpha \in \mathbb{R}$. *Then,*

$$\lambda_{A,B} = \frac{a_{11}+a_{22}}{2} + \frac{1}{2}|a_{12}| \frac{\int\limits_0^\infty \sqrt{v}\exp\left(-\frac{1}{6}|a_{12}|\,v^3+\frac{v}{2|a_{12}|}(\mu_1-\mu_2)^2\right)dv}{\int\limits_0^\infty \frac{1}{\sqrt{v}}\exp\left(-\frac{1}{6}|a_{12}|\,v^3+\frac{v}{2|a_{12}|}(\mu_1-\mu_2)^2\right)dv}.$$

3.3.4.4 Type IV

In this subsection, we consider the case that the matrix B has a pair of conjugated complex eigenvalues, i.e.,

$$B = \begin{pmatrix} \alpha & -\beta \\ \beta & \alpha \end{pmatrix} \qquad \text{where } \alpha, \beta \in \mathbb{R}.$$

Theorem 3.11 (Explicit Formula for the Top Lyapunov Exponents of Linear SDE of Type IV) *Consider system* (3.4) *with the drift part* $A = (a_{ij}) \in \mathbb{R}^{2\times 2}$ *and with the diffusion part* $B = \begin{pmatrix} \alpha & -\beta \\ \beta & \alpha \end{pmatrix}$, *where* $\alpha, \beta \in \mathbb{R}$ *with* $\beta \neq 0$. *Define a function* $\Pi_{A,B} : \left[-\frac{\pi}{2}, \frac{\pi}{2}\right] \to \mathbb{R}$ *by*

$$\Pi_{A,B}(\varphi) := \exp\left(\frac{a_{11}-a_{22}}{2\beta^2}\cos 2\varphi + \frac{a_{12}+a_{21}}{2\beta^2}\sin 2\varphi - \frac{a_{12}-a_{21}}{\beta^2}\varphi\right).$$

Then, the top Lyapunov exponents of (3.4) *are given by:*
If $a_{12} = a_{21}$, *then*

$$\lambda_{A,B} = \frac{a_{11}+a_{22}}{2} + \frac{\int_{-\frac{\pi}{2}}^{\frac{\pi}{2}} \left(\frac{a_{11}-a_{22}}{2}\cos 2\varphi + \frac{a_{12}+a_{21}}{2}\sin 2\varphi\right) \Pi_{A,B}(\varphi)\, d\varphi}{\int_{-\frac{\pi}{2}}^{\frac{\pi}{2}} \Pi_{A,B}(\varphi)\, d\varphi}.$$

If $a_{12} \neq a_{21}$ *then*

$$\lambda_{A,B} = \frac{1}{\Gamma_{A,B}\left(\exp\left(\frac{\pi(a_{12}-a_{21})}{\beta^2}\right)-1\right)} \int_{-\frac{\pi}{2}}^{\frac{\pi}{2}} \frac{1}{\Pi_{A,B}(\varphi)}\, d\varphi \int_{-\frac{\pi}{2}}^{\frac{\pi}{2}} f_A(\varphi)\Pi_{A,B}(\varphi)\, d\varphi$$

$$+\frac{1}{\Gamma_{A,B}} \int_{-\frac{\pi}{2}}^{\frac{\pi}{2}} \int_{-\frac{\pi}{2}}^{\varphi} f_A(\varphi) \frac{\Pi_{A,B}(\varphi)}{\Pi_{A,B}(u)}\, du\, d\varphi.$$

where

$$\Gamma_{A,B} := \int_{-\frac{\pi}{2}}^{\frac{\pi}{2}} \int_{-\frac{\pi}{2}}^{\varphi} \frac{\Pi_{A,B}(\varphi)}{\Pi_{A,B}(u)}\, du\, d\varphi + \frac{\int_{-\frac{\pi}{2}}^{\frac{\pi}{2}} \Pi_{A,B}(\varphi)\, d\varphi \int_{-\frac{\pi}{2}}^{\frac{\pi}{2}} \frac{1}{\Pi_{A,B}(\varphi)}\, d\varphi}{\exp\left(\frac{\pi(a_{12}-a_{21})}{\beta^2}\right)-1}. \tag{3.13}$$

Proof By definition of $g_B(\varphi)$, we have $g_B(\varphi) = \beta$ for all $\varphi \in \left[-\frac{\pi}{2}, \frac{\pi}{2}\right]$. Then, the condition (H) holds for all $A \in \mathbb{R}^{2\times 2}$, and the differential equation of $p(\varphi)$ is given by

$$p'(\varphi) = \frac{2g_A(\varphi)}{\beta^2} p(\varphi) + C, \tag{3.14}$$

where C is a constant. From the definition of $\Pi_{A,B}$, we get that

$$\Pi_{A,B}(\varphi) = \exp\left(\int_{-\frac{\pi}{2}}^{\varphi} \frac{2g_A(u)}{\beta^2}\, du\right) \exp\left(\frac{(a_{12}-a_{21})\pi - (a_{11}-a_{22})}{2\beta^2}\right).$$

Therefore, $\Pi_{A,B}(\varphi)$ is a nontrivial solution of the corresponding linear equation of (3.14) given by $p'(\varphi) = \frac{2g_A(\varphi)}{\beta^2} p(\varphi)$. By variation of constants formula, we arrive at

$$p(\varphi) = \Pi_{A,B}(\varphi)\left(\kappa + C\int_{-\frac{\pi}{2}}^{\varphi} \frac{1}{\Pi_{A,B}(u)}\, du\right), \tag{3.15}$$

where κ is a constant. In the remaining part, we solve κ and C by using the properties that $p(-\frac{\pi}{2}) = p(\frac{\pi}{2})$ and $\int_{-\frac{\pi}{2}}^{\frac{\pi}{2}} p(u)\,du = 1$. From definition of $\Pi_{A,B}$, we have

$$\frac{\Pi_{A,B}(-\frac{\pi}{2})}{\Pi_{A,B}(\frac{\pi}{2})} = \exp\left(\frac{\pi(a_{12}-a_{21})}{\beta^2}\right).$$

So, from (3.15), the equality $p(-\frac{\pi}{2}) = p(\frac{\pi}{2})$ leads to

$$\exp\left(\frac{\pi(a_{12}-a_{21})}{\beta^2}\right)\kappa = \kappa + C\int_{-\frac{\pi}{2}}^{\frac{\pi}{2}} \frac{1}{\Pi_{A,B}(u)}\,du.$$

To solve the preceding equality, we consider the following separated cases:
Case 1: If $a_{12} = a_{21}$, then $C = 0$. Hence, from $\int_{-\frac{\pi}{2}}^{\frac{\pi}{2}} p(\varphi)\,d\varphi = 1$, we derive that $\kappa = \frac{1}{\int_{-\frac{\pi}{2}}^{\frac{\pi}{2}} \Pi_{A,B}(\varphi)\,d\varphi}$. According to Theorem 3.6, we obtain

$$\begin{aligned}\lambda(A,B) &= \frac{\int_{-\frac{\pi}{2}}^{\frac{\pi}{2}} f_A(\varphi)\,\Pi_{A,B}(\varphi)\,d\varphi}{\int_{-\frac{\pi}{2}}^{\frac{\pi}{2}} \Pi_{A,B}(\varphi)\,d\varphi}\\ &= \frac{a_{11}+a_{22}}{2} + \frac{\int_{-\frac{\pi}{2}}^{\frac{\pi}{2}} \left(\frac{a_{11}-a_{22}}{2}\cos 2\varphi + \frac{a_{12}+a_{21}}{2}\sin 2\varphi\right)\Pi_{A,B}(\varphi)\,d\varphi}{\int_{-\frac{\pi}{2}}^{\frac{\pi}{2}} \Pi_{A,B}(\varphi)\,d\varphi}.\end{aligned}$$

Case 2: If $a_{21} \neq a_{12}$, then from $p(-\frac{\pi}{2}) = p(\frac{\pi}{2})$ we derive that

$$\kappa = C\frac{\int_{-\frac{\pi}{2}}^{\frac{\pi}{2}} \frac{1}{\Pi_{A,B}(u)}\,du}{\exp\left(\frac{\pi(a_{12}-a_{21})}{\beta^2}\right)-1}.$$

Since $\int_{-\frac{\pi}{2}}^{\frac{\pi}{2}} p(\varphi)\,d\varphi = 1$ it follows that

$$C\left(\int_{-\frac{\pi}{2}}^{\frac{\pi}{2}}\int_{-\frac{\pi}{2}}^{\varphi} \frac{\Pi_{A,B}(\varphi)}{\Pi_{A,B}(u)}\,du\,d\varphi + \frac{\int_{-\frac{\pi}{2}}^{\frac{\pi}{2}} \Pi_{A,B}(\varphi)\,d\varphi \int_{-\frac{\pi}{2}}^{\frac{\pi}{2}} \frac{1}{\Pi_{A,B}(\varphi)}\,d\varphi}{\exp\left(\frac{\pi(a_{12}-a_{21})}{\beta^2}\right)-1}\right) = 1.$$

Hence, $C = \frac{1}{\Gamma_{A,B}}$, where $\Gamma_{A,B}$ is defined as in (3.13) and using Theorem 3.6, the proof is complete.

3.3.5 Applications

3.3.5.1 Sample-Path Stability of a Linear Stochastic Differential Equation Arsing from Fluid Dynamics

In this subsection, we consider the following linear stochastic differential equation

$$\begin{pmatrix} du_t \\ dv_t \end{pmatrix} = \begin{pmatrix} -R^{-1} & 1 \\ 0 & -R^{-1} \end{pmatrix} \begin{pmatrix} u_t \\ v_t \end{pmatrix} dt + \sigma \begin{pmatrix} 0 & -1 \\ 1 & 0 \end{pmatrix} \begin{pmatrix} u_t \\ v_t \end{pmatrix} \circ dW_t, \tag{3.16}$$

where $\sigma \neq 0$. This system was discussed in [36] and arises from a model in fluid dynamics. In [36], the authors compute explicitly the area of parameters for which system (3.16) is mean-square asymptotically stable. Instead of studying mean-square asymptotic stability, our aim is to find the area of parameters for which system (3.16) is sample-path asymptotically stable.

Theorem 3.12 *System* (3.16) *is asymptotically stable if and only if two parameters* R, σ *satisfy*

$$\frac{\Gamma_{A,B}}{R} > \frac{1}{\exp(\frac{\pi}{\sigma^2}) - 1} \int_{-\frac{\pi}{2}}^{\frac{\pi}{2}} \int_{-\frac{\pi}{2}}^{\frac{\pi}{2}} \frac{\sin 2\varphi}{2} \frac{\exp\left(\frac{\sin 2\varphi}{2\sigma^2} - \frac{\varphi}{\sigma^2}\right)}{\exp\left(\frac{\sin 2u}{2\sigma^2} - \frac{u}{\sigma^2}\right)} du\, d\varphi$$

$$+ \int_{-\frac{\pi}{2}}^{\frac{\pi}{2}} \int_{-\frac{\pi}{2}}^{\varphi} \frac{\sin 2\varphi}{2} \frac{\exp\left(\frac{\sin 2\varphi}{2\sigma^2} - \frac{\varphi}{\sigma^2}\right)}{\exp\left(\frac{\sin 2u}{2\sigma^2} - \frac{u}{\sigma^2}\right)} du\, d\varphi,$$

where

$$\Gamma_{A,B} := \int_{-\frac{\pi}{2}}^{\frac{\pi}{2}} \int_{-\frac{\pi}{2}}^{\varphi} \frac{\exp\left(\frac{\sin 2\varphi}{2\sigma^2} - \frac{\varphi}{\sigma^2}\right)}{\exp\left(\frac{\sin 2u}{2\sigma^2} - \frac{u}{\sigma^2}\right)} du\, d\varphi + \frac{\left(\int_{-\frac{\pi}{2}}^{\frac{\pi}{2}} \exp\left(\frac{\sin 2\varphi}{2\sigma^2} - \frac{\varphi}{\sigma^2}\right) d\varphi\right)^2}{\exp(\frac{\pi}{\sigma^2}) - 1} > 0. \tag{3.17}$$

Proof The diffusion coefficient of this linear stochastic differential equation is of Type IV. Hence, we can use Theorem 3.11 to compute and to determine the sign of the top Lyapunov exponent of (3.16). A direct computation yields that

$$H_{A,B}(\varphi) = \exp\left(\frac{\sin 2\varphi}{2\sigma^2} - \frac{\varphi}{\sigma^2}\right) \quad \text{and} \quad f_A(\varphi) = \frac{1}{R} + \frac{\sin 2\varphi}{2}.$$

Therefore, the constant $\Gamma_{A,B}$ defined in (3.13) is computed explicitly as in (3.17). Hence, by Theorem 3.11, we have

$$\Gamma_{A,B}\lambda_{A,B} = -\frac{1}{R}\Gamma_{A,B} + \frac{1}{\exp(\frac{\pi}{\sigma^2})-1}\int_{-\frac{\pi}{2}}^{\frac{\pi}{2}}\int_{-\frac{\pi}{2}}^{\frac{\pi}{2}} \frac{\sin 2\varphi}{2}\frac{\exp\left(\frac{\sin 2\varphi}{2\sigma^2}-\frac{\varphi}{\sigma^2}\right)}{\exp\left(\frac{\sin 2u}{2\sigma^2}-\frac{u}{\sigma^2}\right)}\,du\,d\varphi$$

$$+\int_{-\frac{\pi}{2}}^{\frac{\pi}{2}}\int_{-\frac{\pi}{2}}^{\varphi} \frac{\sin 2\varphi}{2}\frac{\exp\left(\frac{\sin 2\varphi}{2\sigma^2}-\frac{\varphi}{\sigma^2}\right)}{\exp\left(\frac{\sin 2u}{2\sigma^2}-\frac{u}{\sigma^2}\right)}\,du\,d\varphi,$$

which completes the proof.

As is proved in [36, Theorem 4.2], system (3.16) is mean-square asymptotically stable if R and σ satisfy

$$\frac{6}{R} > -\sigma^2 + \sqrt[3]{8\sigma^6 + 27\sigma^2 + 3\sqrt{3\sigma^4(16\sigma^4+27)}}$$

$$+\frac{4\sigma^4}{\sqrt[3]{8\sigma^6 + 27\sigma^2 + 3\sqrt{3\sigma^4(16\sigma^4+27)}}}.$$

The following lemma is devoted to showing that mean-square asymptotic stability implies sample-path asymptotic stability for an arbitrary planar linear stochastic differential equation

$$\begin{pmatrix} dx_t \\ dy_t \end{pmatrix} = A\begin{pmatrix} x_t \\ y_t \end{pmatrix} dt + B\begin{pmatrix} x_t \\ y_t \end{pmatrix} \circ dW_t. \tag{3.18}$$

Recall that the *mean-square Lyapunov exponent* of (3.18) denoted by $\lambda^{\mathrm{ms}}_{A,B}$ is defined by

$$\lambda^{\mathrm{ms}}_{A,B} := \lim_{t\to\infty}\frac{1}{t}\log \mathrm{e}\|\Phi_{A,B}(t,\cdot)\|^2,$$

see e.g. [9].

Lemma 3.6 *Consider* (3.18) *and suppose that* $\lambda^{\mathrm{ms}}_{A,B} < 0$. *Then,* $\lambda_{A,B} < 0$.

Proof Note that in the case that the coefficients A and B of (3.16) satisfy condition (H) of Theorem 3.6, then the assertion of this lemma was proved in [7, Corollary 1]. Hence, we only need to deal with the case that the pair $\{A, B\}$ does not satisfy (H). Based on the arguments at the beginning of Sect. 3.3.3, we can assume w.l.o.g. that A and B are upper triangular matrices, i.e.,

$$A = \begin{pmatrix} a_{11} & 0 \\ a_{12} & a_{22} \end{pmatrix} \quad \text{and} \quad B = \begin{pmatrix} b_{11} & 0 \\ b_{12} & b_{22} \end{pmatrix}.$$

Using Ito's formula, we obtain that

$$\frac{d\,\mathrm{e}x_t^2}{dt} = 2(a_{11} + b_{11}^2)\,d\mathrm{e}x_t^2,$$

$$\frac{d\,\mathrm{e}x_t y_t}{dt} = \left(a_{11} + a_{22} + \frac{(b_{11} + b_{22})^2}{2}\right)\mathrm{e}x_t y_t,$$

$$+\left(a_{21} + \frac{b_{21}(b_{11} + b_{22})}{2} + b_{11}b_{21}\right)\mathrm{e}x_t^2$$

$$\frac{d\,\mathrm{e}y_t^2}{dt} = 2\left(a_{22} + b_{22}^2\right)\mathrm{e}y_t^2 + b_{21}^2\mathrm{e}x_t^2 + 2\left(a_{21} + \frac{b_{21}(b_{11} + 3b_{22})}{2}\right)\mathrm{e}x_t y_t.$$

Therefore, $\lambda_{A,B}^{\mathrm{ms}} < 0$ is equivalent to the fact that

$$a_{11} + b_{11}^2 < 0 \quad \text{and} \quad a_{22} + b_{22}^2 < 0,$$

which implies together with Theorem 3.7 that $\lambda_{A,B} < 0$. The proof is complete.

3.3.5.2 A Model of Stochastic Hopf Bifurcation with Multiplicative Noise

Consider a model of stochastic Hopf bifurcation with multiplicative noise of the following form

$$\begin{pmatrix} dx_t \\ dy_t \end{pmatrix} = \begin{pmatrix} -y_t + (a - b(x_t^2 + y_t^2))x_t \\ x_t + (a - b(x_t^2 + y_t^2))y_t \end{pmatrix} dt + \begin{pmatrix} \sigma_1 & 0 \\ 0 & \sigma_2 \end{pmatrix}\begin{pmatrix} x_t \\ y_t \end{pmatrix} \circ dW_t, \tag{3.19}$$

where $a \in \mathbb{R}$ and $b, \sigma_1, \sigma_2 > 0$ are parameters. Note that the preceding model with an additional assumption that $\sigma_1 = \sigma_2$ is studied in [21]. Our aim in this section is to find the bifurcation value of (3.19) in the remaining case that $\sigma_1 \neq \sigma_2$. For this purpose, the linearized equation along the trivial solution of (3.19) is

$$\begin{pmatrix} dx_t \\ dy_t \end{pmatrix} = \begin{pmatrix} a & -1 \\ 1 & a \end{pmatrix}\begin{pmatrix} x_t \\ y_t \end{pmatrix} dt + \begin{pmatrix} \sigma_1 & 0 \\ 0 & \sigma_2 \end{pmatrix}\begin{pmatrix} x_t \\ y_t \end{pmatrix} \circ dW_t. \tag{3.20}$$

Let $\delta := \sigma_1 - \sigma_2$, and denote by $a_{\mathrm{bif}}(\delta)$ the bifurcation value satisfying that when a changes increasingly and crosses $a_{\mathrm{bif}}(\delta)$, the system (3.20) changes from stability to instability. For the case that $\delta > 0$, in light of Theorem 3.9, the bifurcation value $a_{\mathrm{bif}}(\delta)$ is

$$a_{\mathrm{bif}}(\delta) = -\frac{\delta^2}{C}\int_0^\infty \frac{\exp\left(\frac{2(1-u^2)}{\delta^2 u}\right)}{1+u^2}\int_0^u \frac{\exp\left(-\frac{2(1-v^2)}{\delta^2 v}\right)}{1+v^2}\,dv\,du$$

$$-\frac{\delta^2}{C}\int_{-\infty}^{0}\frac{\exp\left(\frac{2(1-u^2)}{\delta^2 u}\right)}{1+u^2}\int_{-\infty}^{u}\frac{\exp\left(-\frac{2(1-v^2)}{\delta^2 v}\right)}{1+v^2}\,dv\,du,$$

where

$$C := \int_0^{\infty}\frac{1+u^2}{u^2}\exp\left(\frac{2(1-u^2)}{\delta^2 u}\right)\int_0^{u}\frac{\exp\left(\frac{2(1-v^2)}{\delta^2 v}\right)}{1+v^2}\,dv\,du$$

$$+\int_{-\infty}^{0}\frac{1+u^2}{u^2}\exp\left(\frac{2(1-u^2)}{\delta^2 u}\right)\int_{-\infty}^{u}\frac{\exp\left(\frac{2(1-v^2)}{\delta^2 v}\right)}{1+v^2}\,dv\,du.$$

By virtue of Theorem 3.8, $a_{\text{bif}}(0) = 0$, i.e., when $\delta = 0$ the bifurcation value of the stochastic Hopf bifurcation model (3.19) coincides with the Hopf bifurcation value of the deterministic part. Meanwhile, when $\delta > 0$, the bifurcation value of the stochastic Hopf bifurcation model (3.19) is strictly smaller than the Hopf bifurcation value of the deterministic part.

Chapter 4
Genericity of Lyapunov Spectrum of Random Dynamical Systems

Abstract Our first aim in this chapter is to prove genericity of analyticity of Lyapunov spectrum for discrete-time bounded linear random dynamical systems and bounded linear random differential equations. The main ingredient in the proof is the openness and denseness of integral separations for these bounded linear random dynamical systems. This property is no longer true for unbounded linear random dynamical systems. So, our second aim in this chapter is to construct an open set of unbounded linear random dynamical systems with simple Lyapunov spectrum but having no exponential separated splitting. This example leads to several open questions on the genericity of Lyapunov spectrum of unbounded linear random dynamical systems. The materials of this chapter are taken from Cong and Doan (Stoch Dyn. 7(3):335–355, 2007), (Discrete Contin Dyn Syst Seri S 9:995–1007, 2016), Doan (Discrete Contin Dyn Syst B 22(8):3113–3126, 2017).

4.1 The Spaces of Linear Random Dynamical Systems

In this section, we introduce two spaces of linear random dynamical systems (discrete and continuous-time).

4.1.1 *Discrete-Time Bounded Linear Random Dynamical Systems*

Let $(X, \mathcal{A}, m)$ be a Lebesgue probability space and $T : X \to X$ an invertible ergodic transformation preserving the probability m. Let $\mathcal{L}^\infty(X, \mathbb{R}^{d\times d})$ denotes the space of bounded measurable mappings from X to $\mathbb{R}^{d\times d}$. Each random map $A \in \mathcal{L}^\infty(X, \mathbb{R}^{d\times d})$ gives rise to an one-sided linear random dynamical system[1] Φ_A :

[1] In this chapter, we identify a random map in $\mathcal{L}^\infty(X, \mathbb{R}^{d\times d})$ with its generated linear random dynamical system.

T. S. Doan, *Spectral Theory of Nonautonomous Dynamical Systems and Applications*, https://doi.org/10.1007/978-981-97-5520-2_4

$\mathbb{Z}_{\geq 0} \times X \to \mathbb{R}^{d\times d}$ via

$$\Phi_A(n,x) := \begin{cases} A(T^{n-1}x)\dots A(x), & \text{if } n > 0, \\ \text{id}, & \text{if } n = 0, \end{cases}$$

where id denotes the identity matrix in $\mathbb{R}^{d\times d}$. By virtue of the multiplicative ergodic theorem for one-sided linear random dynamical systems defined on an invertible metric dynamical system, see, e.g., [65]; there exist finite numbers of Lyapunov exponents $\lambda_1(A) > \cdots > \lambda_{p(A)}(A)$, where $p(A) \in \{1,\dots,d\}$, and a forward invariant decomposition

$$\mathbb{R}^d = O_1(A,x) \oplus O_2(A,x) \oplus \cdots \oplus O_{p(A)}(A,x)$$

with the property that

$$\lim_{n\to\infty} \frac{1}{n} \log \|\Phi_A(n,x)v\| = \lambda_i \quad \text{iff} \quad v \in O_i(A,x) \setminus \{0\}.$$

Let $d_1(A),\dots,d_{p(A)}(A)$ denote the dimension of $O_1(A,x),\dots,O_{p(A)}(A,x)$, respectively. The set of pairs

$$\Sigma_{\text{Lya}}(A) := \big\{(\lambda_1(A), d_1(A)),\dots,(\lambda_{p(A)}, d_{p(A)}(A))\big\}$$

is called the *Lyapunov spectrum* of A. For a convenience of presentation in the rest of the paper, let $\gamma_d(A) \leq \gamma_{d-1}(A) \leq \cdots \leq \gamma_1(A)$ denote the Lyapunov exponents of A (counting with multiplicity), i.e., for $i = 1,\dots,p(A)$

$$\gamma_j(A) = \lambda_i(A) \qquad \text{for } d_1(A) + \cdots + d_{i-1}(A) + 1 \leq j \leq d_1(A) + \cdots + d_i(A).$$

The Lyapunov spectrum is said to be *simple* if $p(A) = d$, i.e., all Oseledets subspaces $O_1(A,x),\dots,O_{p(A)}(A,x)$ are of dimension 1.

Let $\text{GL}(d)$ be the set of all invertible matrices in $\mathbb{R}^{d\times d}$. Let $\mathcal{L}^\infty(X,\text{GL}(d))$ denote the space of all $A \in \mathcal{L}^\infty(X,\mathbb{R}^{d\times d})$ such that the inverse $A^{-1}(x)$ exists for $x \in X$ and $A^{-1}(\cdot) \in \mathcal{L}^\infty(X,\mathbb{R}^{d\times d})$. Note that each random map $A \in \mathcal{L}^\infty(X,\text{GL}(d))$ generates a two-sided linear random dynamical system which is also denoted by $\Phi_A : \mathbb{Z} \times X \to \text{GL}(d)$ via

$$\Phi_A(n,x) := \begin{cases} A(T^{n-1}x)\dots A(x), & \text{if } n > 0, \\ \text{id}, & \text{if } n = 0, \\ A(T^n x)^{-1} \circ \cdots \circ \mathcal{A}(T^{-1}x)^{-1}, & \text{if } n < 0. \end{cases}$$

In the following lemma, we show that $\mathcal{L}^\infty(X,\text{GL}(d))$ is an open and dense subset of $\mathcal{L}^\infty(X,\mathbb{R}^{d\times d})$. Note that this result does not depend on the norm equipped with $\mathbb{R}^d$ and the induced norm on $\mathbb{R}^{d\times d}$. Then, in what follows let $\mathbb{R}^d$ be endowed with the standard Euclidean norm.

Lemma 4.1 *The set $\mathcal{L}^\infty(X, \mathrm{GL}(d))$ is open and dense in the Banach space $(\mathcal{L}^\infty(X, \mathbb{R}^{d\times d}), \|\cdot\|_\infty)$.*

Proof The openness of $\mathcal{L}^\infty(X, \mathrm{GL}(d))$ in $(\mathcal{L}^\infty(X, \mathbb{R}^{d\times d}), \|\cdot\|_\infty)$ is obvious, and we only need to prove the density of this set. Let $A \in \mathcal{L}^\infty(X, \mathbb{R}^{d\times d})$ and $\varepsilon > 0$ be arbitrary. Let $A(x) = U(x)\Sigma(x)V(x)$ be a measurable singular valued decomposition of A, where $U(x)$ and $V(x)$ are measurable orthogonal matrices and $\Sigma(x) = \mathrm{diag}(\delta_1(x), \ldots, \delta_d(x))$ is a diagonal matrix. Define $\widehat{A}(x) := U(x)\widehat{\Sigma}(x)V(x)$, where $\widehat{\Sigma}(x) := \mathrm{diag}(\widehat{\delta_1}(x), \ldots, \widehat{\delta_d}(x))$ is defined by

$$\widehat{\delta_i}(x) := \begin{cases} \delta_i(x), & \text{if } |\delta_i(x)| \geq \frac{\varepsilon}{2}, \\ \frac{\varepsilon}{2}, & \text{if } |\delta_i(x)| < \frac{\varepsilon}{2}. \end{cases}$$

Then, $\|A - \widehat{A}\|_\infty = \mathrm{ess\,sup}_{x\in X} \|\Sigma(x) - \widehat{\Sigma}(x)\| \leq \varepsilon$. Furthermore, $\widehat{A}(x)$ is invertible and $\|\widehat{A}^{-1}\|_\infty = \mathrm{ess\,sup}_{x\in X} \|\widehat{A}^{-1}(x)\| \leq \frac{2}{\varepsilon}$. Hence, $\mathcal{L}^\infty(X, \mathrm{GL}(d))$ is dense in $(\mathcal{L}^\infty(X, \mathbb{R}^{d\times d}), \|\cdot\|_\infty)$, and the proof is complete.

4.1.2 Bounded Linear Random Differential Equations

Let $(\Omega, \mathcal{F}, \mathbb{P})$ be a Lebesgue probability space and $(\theta_t)_{t\in\mathbb{R}}$ be an ergodic flow from Ω into itself preserving the probability $\mathbb{P}$. Suppose further that $(\theta_t)_{t\in\mathbb{R}}$ has no fixed point. The assumption of ergodicity of the flow is natural, because one may easily use ergodic decomposition of Ω (see [40]) to reduce the problem to the ergodic case. Furthermore, we do not consider the fixed point of $(\theta_t)_{t\in\mathbb{R}}$, because this is the case of trivial dynamics, and our problem reduces to the well-known problem of autonomous linear ordinary differential equations.

Let $\mathcal{L}^\infty(\Omega, \mathbb{R}^{d\times d})$ denote the space of bounded measurable matrix-valued maps $A : \Omega \to \mathbb{R}^{d\times d}$ satisfying that

$$\operatorname*{ess\,sup}_{\omega\in\Omega} \|A(\omega)\| < \infty.$$

We endow $\mathcal{L}^\infty(\Omega, \mathbb{R}^{d\times d})$ with the L^∞-metric ρ_∞ defined by

$$\rho_\infty(A, B) := \operatorname*{ess\,sup}_{\omega\in\Omega} \|A(\omega) - B(\omega)\|.$$

It is well known that the metric space $(\mathcal{L}^\infty(\Omega, \mathbb{R}^{d\times d}), \rho_\infty)$ is complete; see, e.g., [8]. For each $A \in \mathcal{L}^\infty(\Omega, \mathbb{R}^{d\times d})$, we consider the corresponding *linear random differential equation* of the following form (see [6, Subsection 2.2])

$$\dot{x} = A(\theta_t \omega)x \qquad \text{for } \omega \in \Omega, t \in \mathbb{R}. \tag{4.1}$$

Let $\Psi_A(t, \omega)x$ denote the solution of (4.1) satisfying $x(0) = x$, i.e., $\Psi_A(t, \omega)$ satisfies the following integral equation

$$\Psi_A(t, \omega) = \text{id} + \int_0^t A(\theta_s \omega)\Psi_A(s, \omega)\, ds \quad \text{for all } t \in \mathbb{R}, \tag{4.2}$$

where id denotes the identity matrix of size d. Thus, the linear mapping $\Psi_A : \mathbb{R} \times \Omega \to \mathbb{R}^{d\times d}$ is a continuous random dynamical system, i.e., Ψ_A is $(\mathcal{B}(\mathbb{R}) \otimes \mathcal{F}, \mathcal{B}(\mathbb{R}^{d\times d}))$ measurable and the following properties hold:

(i) $\Psi_A(0, \omega) = \text{id}$,
(ii) $\Psi_A(t + s, \omega) = \Phi_A(t, \theta_s \omega)\Phi_A(s, \omega)$ for all $t, s \in \mathbb{R}, \omega \in \Omega$,
(iii) For each $\omega \in \Omega$, the mapping $t \mapsto \Psi_A(t, \omega)$ is continuous,

see, e.g., [6, Subsection 2.2] and Definition 3.3. In order to provide some estimates on $\|\Psi_A(t, \omega)\|$, we need the following technical lemma. This lemma can be considered as a L^∞-version of [6, Lemma 2.2.5].

Lemma 4.2 *Let $A \in \mathcal{L}^\infty(\Omega, \mathbb{R}^{d\times d})$ and define $K := \operatorname{ess\,sup}_{\omega\in\Omega} \|A(\omega)\|$. Then, for any $\varepsilon > 0$, there exists a measurable set $\widehat{\Omega} \subset \Omega$ (depending on ε) being of full measure and invariant under the flow $(\theta_t)_{t\in\mathbb{R}}$, i.e., $\theta_t\widehat{\Omega} \subset \widehat{\Omega}$ for all $t \in \mathbb{R}$, such that for $\alpha, \beta \in \mathbb{R}$ with $\alpha \le \beta$*

$$\int_\alpha^\beta \|A(\theta_s \omega)\|\, ds \le (K + \varepsilon)(\beta - \alpha) \quad \textit{for all } \omega \in \widehat{\Omega}. \tag{4.3}$$

Proof Choose and fix an arbitrary $\varepsilon > 0$. Let $\ell \in \mathbb{Q}_{>0}$ be an arbitrary positive rational number. Define the following set

$$\mathcal{M}_\ell := \left\{\omega \in \Omega : \int_a^b \|A(\theta_s \omega)\|\, ds \le (K + \varepsilon)\ell \quad \text{for all } b - a = \ell\right\}. \tag{4.4}$$

By continuity of Lebesgue integral, we have

$$\mathcal{M}_\ell := \bigcap_{a\in\mathbb{Q}} \left\{\omega \in \Omega : \int_a^{a+\ell} \|A(\theta_s \omega)\|\, ds \le (K + \varepsilon)\ell\right\},$$

which implies that the set $\mathcal{M}_\ell$ is measurable. From the following observation

$$\int_a^b \|A(\theta_s \circ \theta_t \omega)\|\, ds = \int_{a+t}^{b+t} \|A(\theta_s \omega)\|\, ds \qquad \text{for all } t \in \mathbb{R},$$

we derive that

$$\theta_t \mathcal{M}_\ell \subset \mathcal{M}_\ell \qquad \text{for all } t \in \mathbb{R}.$$

Thus, by ergodicity of the flow $(\theta_t)_{t\in\mathbb{R}}$, either $\mathbb{P}(\mathcal{M}_\ell) = 0$ or $\mathbb{P}(\mathcal{M}_\ell) = 1$ holds. We now prove that $\mathbb{P}(\mathcal{M}_\ell) = 1$ by supposing the contrary that $\mathbb{P}(\mathcal{M}_\ell) = 0$. Let $\delta := \frac{\varepsilon\ell}{2K} > 0$, and for each $k \in \mathbb{Z}$, we define

$$\mathcal{N}_{k,\ell} := \left\{\omega \in \Omega : \int_a^{a+\ell} \|A(\theta_s\omega)\|\, ds \geq (K+\varepsilon)\ell \text{ for some } a \in [k\delta, (k+1)\delta]\right\}.$$

Using analogous arguments as in the proof of measurability of $\mathcal{M}_\ell$ above, the set $\mathcal{N}_{k,\ell}$ is also measurable. By definition of $\mathcal{M}_\ell$ as in (4.4), we have $\mathcal{M}_\ell^c = \bigcup_{k\in\mathbb{Z}} \mathcal{N}_{k,\ell}$, where $\mathcal{M}_\ell^c$ is the complement of $\mathcal{M}_\ell$. Thus, $\mathbb{P}(\mathcal{M}_\ell^c) = 1$ implies that there exists $k \in \mathbb{Z}$ such that $\mathbb{P}(\mathcal{N}_{k,\ell}) > 0$. Note that $\mathcal{N}_{k,\ell}$ is a subset of the following set

$$\mathcal{N} := \left\{\omega \in \Omega : \int_{k\delta}^{(k+1)\delta+\ell} \|A(\theta_s\omega)\|\, ds \geq (K+\varepsilon)\ell\right\}. \tag{4.5}$$

By measurability of the maps $(s, \omega) \mapsto \theta_s\omega$ and $\omega \mapsto A(\omega)$, we have that the map $(s, \omega) \mapsto \|A(\theta_s\omega)\|$ is also measurable. Thus, using Fubini's theorem, we obtain that

$$\int_{k\delta}^{(k+1)\delta+\ell} \int_{\mathcal{N}} \|A(\theta_s\omega)\|\, d\mathbb{P}(\omega)\, ds = \int_{\mathcal{N}} \int_{k\delta}^{(k+1)\delta+\ell} \|A(\theta_s\omega)\|\, ds\, d\mathbb{P}(\omega),$$

which together with (4.5) implies that

$$\int_{k\delta}^{(k+1)\delta+\ell} \int_{\mathcal{N}} \|A(\theta_s\omega)\|\, d\mathbb{P}(\omega)\, ds \geq (K+\varepsilon)\ell\ \mathbb{P}(\mathcal{N}). \tag{4.6}$$

From the fact that $K = \operatorname{ess\,sup}_{\omega\in\Omega} \|A(\omega)\|$ we derive that

$$\int_{\mathcal{N}} \|A(\theta_s\omega)\|\, d\mathbb{P}(\omega) \leq K\ \mathbb{P}(\mathcal{N}) \quad \text{for all } s \in [k\delta, (k+1)\delta+\ell],$$

which gives

$$\int_{k\delta}^{(k+1)\delta+\ell} \int_{\mathcal{N}} \|A(\theta_s\omega)\|\, d\mathbb{P}(\omega)\, ds \leq (\delta+\ell)K\ \mathbb{P}(\mathcal{N}).$$

Since $\delta = \frac{\varepsilon\ell}{2K}$, it follows that $(\delta+\ell)K < (K+\varepsilon)\ell$. Thus, the preceding inequality provides a contradiction to (4.6). Therefore, $\mathbb{P}(\mathcal{M}_\ell) = 1$ for all $\ell \in \mathbb{Q}_{\geq 0}$. Hence, the measurable set $\widehat{\Omega} := \bigcap_{\ell\in\mathbb{Q}_{>0}} \mathcal{M}_\ell$ is of full measure and also invariant under all θ_t for $t \in \mathbb{R}$. To conclude the proof, we prove that $\widehat{\Omega}$ satisfies (4.3). For this purpose, let $\omega \in \widehat{\Omega}$ and $\alpha, \beta \in \mathbb{R}$ with $\alpha < \beta$. Let $(q_n)_{n\in\mathbb{N}}$ be a sequence of positive rational numbers such that $\lim_{n\to\infty} q_n = \beta - \alpha$. Since $\omega \in \widehat{\Omega}$, it follows that $\omega \in \mathcal{M}_{q_n}$ for all $n \in \mathbb{N}$. Therefore,

$$\int_{\alpha}^{\alpha+q_n} \|A(\theta_s\omega)\| \, ds \le (K+\varepsilon)q_n.$$

Letting $n \to \infty$ in the above inequality, we obtain that

$$\int_{\alpha}^{\beta} \|A(\theta_s\omega)\| \, ds \le (K+\varepsilon)(\beta-\alpha),$$

which completes the proof.

According to Lemma 4.2, replacing Ω by a set of full measure being invariant under $(\theta_t)_{t\in\mathbb{R}}$, we can assume from now on that

$$\int_{\alpha}^{\beta} \|A(\theta_s\omega)\| \, ds \le \hat{K}(\beta-\alpha) \quad \text{for all } \omega \in \Omega, \alpha, \beta \in \mathbb{R}, \tag{4.7}$$

where $\hat{K} > \operatorname{ess\,sup}_{\omega\in\Omega} \|A(\omega)\|$ is a positive constant. In the following corollary, we establish some estimates on the norm of the solution $\Psi_A(t,\omega)$.

Corollary 4.1 *For all $\omega \in \Omega$ and $t \in \mathbb{R}$, we have*

$$\|\Psi_A(t,\omega)\| \le \exp\left(|t|\hat{K}\right),$$

where $\hat{K}$ is a positive constant satisfying (4.7).

Proof Choose and fix $\omega \in \Omega$. Let $\delta > 0$ be arbitrary such that $\hat{K}\delta < 1$. From (4.2), we obtain that

$$\|\Psi_A(\tau,\omega)\| \le 1 + \left|\int_0^{\tau} \|A(\theta_s\omega)\| \|\Psi_A(s,\omega)\| \, ds\right| \quad \text{for } \tau \in [0,\delta],$$

which together with (4.7) implies that

$$\max_{0\le s\le\delta} \|\Psi_A(s,\omega)\| \le 1 + \hat{K}\delta \max_{0\le s\le\delta} \|\Psi_A(s,\omega)\|.$$

Note that the maximum $\max_{0\le s\le\delta} \|\Psi_A(s,\omega)\|$ exists due to continuity of the map $s \mapsto \Psi_A(s,\omega)$. Since $\hat{K}\delta < 1$, it follows that

$$\max_{0\le s\le\delta} \|\Psi_A(s,\omega)\| \le \frac{1}{1-\hat{K}\delta}. \tag{4.8}$$

Now let $t > 0$ be arbitrary. We choose $n \in \mathbb{N}$ such that $\hat{K}\frac{t}{n} < 1$. By cocycle property, we have

$$\Psi_A(t, \omega) = \prod_{k=0}^{n-1} \Psi_A\left(\frac{t}{n}, \theta_{k\frac{t}{n}}\omega\right).$$

Hence, using (4.8), we obtain

$$\|\Psi_A(t, \omega)\| \leq \left(\frac{1}{1 - \hat{K}\frac{t}{n}}\right)^n.$$

Letting $n \to \infty$ yields that $\|\Psi_A(t, \omega)\| \leq \exp(\hat{K}t)$. Similarly, for the case $t < 0$, we also obtain that $\|\Psi_A(t, \omega)\| \leq \exp(-\hat{K}t)$. The proof is complete.

Using Corollary 4.1, the generated random dynamical system $\Phi_A(t, \omega)$ satisfies the integrability condition of the multiplicative ergodic theorem, i.e.,

$$\ln^+ \|\alpha^+(\cdot)\|, \ln^+ \|\alpha^-(\cdot)\| \in \mathcal{L}^1(\Omega, \mathcal{F}, \mathbb{P}),$$

where $\ln^+ x := \max(0, \ln x)$ and

$$\alpha^+(\omega) := \sup_{0 \leq t \leq 1} \|\Psi_A(t, \omega)\|, \alpha^-(\omega) := \sup_{-1 \leq t \leq 0} \|\Psi_A(t, \omega)\|.$$

So, we arrive at the following result on the asymptotic behavior of the solution $\Psi_A(t, \omega)v$ for an arbitrary $v \in \mathbb{R}^d \setminus \{0\}$ as $t \to \pm\infty$.

Theorem 4.1 (Multiplicative Ergodic Theorem for Linear Random Differential Equations) *Let Ψ_A denote the linear random dynamical system generated by* (4.1). *Then, there exist p, where $1 \leq p \leq d$, nonrandom Lyapunov exponents $\lambda_p < \lambda_{p-1} < \cdots < \lambda_1$ and an invariant measurable decomposition*

$$\mathbb{R}^d = O_1(A, \omega) \oplus O_2(A, \omega) \oplus \cdots \oplus O_p(A, \omega)$$

with the property that for $k = 1, \ldots, p$, the linear measurable subspace $O_k(\omega)$ is dynamically characterized by

$$\lim_{t \to \pm\infty} \frac{1}{t} \ln \|\Psi_A(t, \omega)v\| = \lambda_k \quad \textit{iff} \quad v \in O_k(A, \omega) \setminus \{0\}.$$

Proof See [6, Theorem 3.4.11].

Analog to the case of discrete-time bounded linear random dynamical systems, let $\gamma_d(A) \leq \gamma_{d-1}(A) \leq \cdots \leq \gamma_1(A)$ denote the Lyapunov exponents (counting with multiplicity) of the linear random dynamical system Ψ_A generated by (4.1).

4.2 Integrally Separated Linear Random Dynamical Systems

4.2.1 Exponential Separation and Integral Separation of Discrete-Time Bounded Linear Random Dynamical Systems

We start by giving the notion of *exponential separation*. Note that is equivalent to the notion of *domination* introduced by Bochi and Viana [26, 27] for classical dynamical systems on compact manifolds and cocycles over them. It is also a random version of the notion of exponential separation of ordinary differential equations which originated from the works of Bylov, Coppel, Sacker and Sell, Palmer, and others (see [37, 47, 100, 101, 108] and Definition 1.5). For linear cocycles, there is also a notion of *integral separateness* introduced by Nguyen Dinh Cong [43] which is a random version of the notion of integral separateness of linear systems of differential equations and is equivalent to the notion of exponential separation (see Bylov et al. [37] for the case of ordinary differential equations). Although these terms are equivalent in a sense, for linear cocycles, we prefer the term *exponential separation* which has longer history and emphasizes the "separation" of invariant subspaces.

Definition 4.1 (Exponential Separation and Integral Separation of Discrete-Time Bounded Linear Random Dynamical Systems) Let $A \in \mathcal{L}^\infty(X, \mathrm{GL}(d))$ and

$$\mathbb{R}^d = E_1(x) \oplus \cdots \oplus E_k(x), \quad k \geqslant 2, \tag{4.9}$$

be an invariant splitting of A, i.e., for almost all $x \in X$ and any $i = 1, \ldots, k$, we have $A(x)E_i(x) = E_i(Tx)$. The splitting (4.9) is called *exponentially separated* if there exist positive numbers $\alpha, K > 0$ such that for any $n \in \mathbb{N}$, $\omega \in \Omega$, and any $i = 1, \ldots, k-1$, the inequality

$$\frac{\|\Phi_A(n,x)u\|}{\|u\|} \leqslant Ke^{-n\alpha} \cdot \frac{\|\Phi_A(n,x)v\|}{\|v\|}$$

holds for all $0 \neq u \in E_1(x) \oplus \cdots \oplus E_i(x)$ and $0 \neq v \in E_{i+1}(x) \oplus \cdots \oplus E_k(x)$. When A has an exponential splitting

$$\mathbb{R}^d = E_1(x) \oplus \cdots \oplus E_d(x)$$

with $\dim E_1(x) = \cdots = \dim E_d(x) = 1$, then A is said to be *integrally separated.*

Now, we give a comparison between exponential separation and exponential dichotomy. Firstly, like exponential dichotomy, in case of bounded cocycle, exponential separation implies boundedness from below of the angle between invariant subspaces of the splitting (see also Bochi and Viana [26]).

Proposition 4.1 *If $A \in \mathcal{L}^\infty(X, \mathrm{GL}(d))$ has an exponentially separated splitting*

$$\mathbb{R}^d = E_1(x) \oplus \cdots \oplus E_k(x),$$

then for any $i = 1, \ldots, k-1$, the angle between $E_i(x)$ and $E_1(x) \oplus \cdots \oplus E_{i-1}(x) \oplus E_{i+1}(x) \oplus \cdots \oplus E_k(x)$ is separated from zero by a positive constant independent of $\omega \in \Omega$.

Proof Put $F_i(x) := E_1(x) \oplus \cdots \oplus E_i(x)$ and $G_{i+1}(x) := E_{i+1}(x) \oplus \cdots \oplus E_k(x)$. Further, $M := \operatorname*{ess\,sup}_{x\in X} \|A(x)\| + \operatorname*{ess\,sup}_{x\in X} \|A^{-1}(x)\| < \infty$. Let m be a positive integer such that $2Ke^{-m\alpha} < 1$, where K, α are the constants provided by the definition of exponential separation of A. Let $u \in F_i(x)$ and $v \in G_{i+1}(x)$ be arbitrary unit vectors. We have

$$2\|\Phi_A(m, x)u\| \leqslant \|\Phi_A(m, x)v\| \leqslant \|\Phi_A(m, x)u\| + \|\Phi_A(m, x)(v - u)\|,$$

hence $\|\Phi_A(m, x)u\| \leqslant \|\Phi_A(m, x)(v - u)\| \leqslant \|\Phi_A(m, x)\| \|v - u\|$. Consequently,

$$\|v - u\| \geqslant \|\Phi_A(m, x)\|^{-1} \|\Phi_A(m, x)^{-1}\|^{-1} \geqslant M^{-2m},$$

which implies $\measuredangle(u, v) \geqslant M^{-2m}$ and the angle between $F_i(x)$ and $G_{i+1}(x)$ is separated from zero by $M^{-2m} > 0$.

Now, let $z \in E_i(x)$ and $h \in F_{i-1}(x) \oplus G_{i+1}(x)$ be arbitrary unit vectors. Then $h = a_1 u + a_2 v$ for some $u \in F_{i-1}(x)$, $v \in G_{i+1}(x)$, $a_1, a_2 \in \mathbb{R}$, $\|u\| = \|v\| = 1$. Clearly, $\max\{|a_1|, |a_2|\} \geqslant 1/2$. Suppose $|a_1| \geqslant 1/2$, setting $r := \frac{a_2}{a_1} v - \frac{1}{a_1} z$ we have

$$\|h - z\| = \|a_1 u + a_2 v - z\| = |a_1| \|u + \frac{a_2}{a_1} v - \frac{1}{a_1} z\| = |a_1| \|u + r\|.$$

Since $\measuredangle(F_{i-1}(x), G_i(x)) \geqslant M^{-2m}$, $r \in G_i(x)$, $u \in F_{i-1}(x)$, $\|u\| = 1$, we have $\|u + r\| \geqslant \sin M^{-2m} \geqslant M^{-2m}/2$, hence $\|h - z\| \geqslant M^{-2m}/4$. Similarly for the case $|a_2| \geqslant 1/2$. Thus, $\measuredangle(E_i(x), F_{i-1}(x) \oplus G_{i+1}(x)) \geqslant M^{-2m}/4$.

The property of having an exponentially separated splitting is also robust.

Proposition 4.2 *If $A \in \mathcal{L}^\infty(X, \mathrm{GL}(d))$ has an exponentially separated splitting then there exists $\varepsilon > 0$ such that any cocycle $B \in \mathcal{L}^\infty(X, \mathrm{GL}(d))$ satisfying $\operatorname*{ess\,sup}_{x\in X} \|A(x) - B(x)\| < \varepsilon$ also has an exponentially separated splitting. Moreover, for small ε, the constants and the splitting of the exponential separation B are close to those of A.*

The proof of this proposition is analogous to that of robustness of exponential dichotomy (see also Bochi and Viana [26]).

Corollary 4.2 *If $A \in \mathcal{L}^\infty(X, \mathrm{GL}(d))$ has an exponentially separated splitting, then there exists $\varepsilon > 0$ such that any cocycle $B \in \mathcal{L}^\infty(X, \mathrm{GL}(d))$ satisfying*

$$\operatorname*{ess\,sup}_{x\in X} \|A(x) - B(x)\| + \operatorname*{ess\,sup}_{x\in X} \|A(x)^{-1} - B(x)^{-1}\| < \varepsilon$$

also has an exponentially separated splitting.

Remark 4.1 Corollary 4.2 is in fact equivalent to Proposition 4.2 because in $\mathcal{L}^\infty(X, \mathrm{GL}(d))$, the distance defined by $\operatorname*{ess\,sup}_{x\in X} \|A(x) - B(x)\|$ is equivalent to the one defined by $\operatorname*{ess\,sup}_{x\in X} \|A(x) - B(x)\| + \operatorname*{ess\,sup}_{x\in X} \|A(x)^{-1} - B(x)^{-1}\|$. To see this, use the formula $B(x)^{-1} - A(x)^{-1} = B(x)^{-1}(A(x) - B(x))A(x)^{-1}$.

Now, we show that exponential dichotomy is strictly stronger than exponential separation

Theorem 4.2 *Suppose that $A \in \mathcal{L}^\infty(X, \mathrm{GL}(d))$ exhibits an exponential dichotomy with constants $K > 0, \alpha > 0$ and a family of projections $P(x)$ of $\mathbb{R}^d$. Then the invariant splitting*

$$\mathbb{R}^d = E_1(x) \oplus E_2(x),$$

where $E_1(x) := P(x)\mathbb{R}^d$ and $E_2(x) := (id - P(x))\mathbb{R}^d$, is exponentially separated.

Proof Note that the spaces $E_1(x)$ and $E_2(x)$ are invariant with respect to A. For any nonvanishing vectors $u \in E_1(x)$ and $v \in E_2(x)$ and $n \in \mathbb{N}$, we have

$$\frac{\|\Phi_A(n, x)u\|}{\|u\|} \leqslant \|\Phi_A(n, x)P(x)\| \leqslant Ke^{-\alpha n},$$

$$\frac{\|\Phi_A(n, x)v\|}{\|v\|} \geqslant \frac{1}{\|(I - P(x))\Phi_A(n, x)^{-1}\|} \geqslant \frac{e^{\alpha n}}{K}.$$

Therefore, the splitting $\mathbb{R}^d = E_1(x) \oplus E_2(x)$ is exponentially separated.

The following proposition (based on a one-dimensional example by Nguyen Dinh Cong [42, Proposition 3.2]) shows us a bounded linear two-dimensional cocycle which has an exponentially separated splitting but does not exhibit exponential dichotomy. First we recall from [42] a technical lemma which will be needed later.

Lemma 4.3 *There exists a measurable set $U \subset X$ which can be represented in the form*

$$U = \bigcup_{k=4}^{\infty} \bigcup_{j=0}^{3k-1} T^j U_k, \tag{4.10}$$

where the sets $T^j U_k$, $k = 0, 1, \ldots, j = 0, \ldots, 3k - 1$, are pairwise disjoint and are all of positive m-measure.

Proposition 4.3 *There exists* $A \in \mathcal{L}^\infty(X, \mathrm{GL}(2))$ *and* $0 < \varepsilon < 1$ *such that any cocycle* $B \in L^\infty(X, \mathrm{GL}(2))$ *satisfying*

$$\operatorname*{ess\,sup}_{x\in X} \|A(x) - B(x)\| + \operatorname*{ess\,sup}_{x\in X} \|A(x)^{-1} - B(x)^{-1}\| < \varepsilon$$

is hyperbolic and the Oseledets splitting of B is exponentially separated, but B exhibits no exponential dichotomy.

Proof Let $U \subset X$ be a set with representation (4.10) provided by Lemma 4.3. It is easily seen that we may assume $m(U) < \frac{1}{4}$. We construct a cocycle

$$A(x) := \begin{pmatrix} a_1(x) & 0 \\ 0 & a_2(x) \end{pmatrix} \in \mathcal{L}^\infty(X, \mathrm{GL}(2))$$

by setting

$$a_1(x) = \begin{cases} \frac{1}{4} & \text{for } x \in U, \\ 1 & \text{for } x \in X \setminus U, \end{cases}$$

$$a_2(x) = \begin{cases} \frac{1}{2} & \text{for } \omega \in U, \\ 2 & \text{for } \omega \in \Omega \setminus U. \end{cases}$$

Clearly,

$$\mathbb{R}^2 = \mathbb{R}e_1 \oplus \mathbb{R}e_2 \tag{4.11}$$

is an invariant and exponentially separated splitting of A. The Lyapunov exponents of A are easily computed, and since $m(U) < 1/4$, we have

$$\lambda_1(A) = \int_X \log |a_1(x)| \, dm(x) < 0,$$

$$\lambda_2(A) = \int_X \log |a_2(x)| \, dm(x) > 0,$$

hence A is hyperbolic. By Proposition 4.2, there is $\varepsilon > 0$ such that any $B \in \mathcal{L}^\infty(X, \mathrm{GL}(2))$ satisfying $\operatorname*{ess\,sup}_{x\in X} \|A(x) - B(x)\| + \operatorname*{ess\,sup}_{x\in X} \|A(x)^{-1} - B(x)^{-1}\| < \varepsilon$ has an exponentially separated splitting which is close to the splitting (4.11). By making $\varepsilon > 0$ smaller if necessary we can show that B is also hyperbolic. Note that along the orbit segments on U, the norm of $A(x)$ equals 1/2; hence, for $\varepsilon < 1/4$, we have $\|B(x)\| < 3/4 < 1$. The set U contains arbitrarily long orbit segments, and on these segments, there are no expanding direction for B. Hence B has no uniformly expanding direction, and thus it cannot exhibit exponential dichotomy.

Remark 4.2

(i) A higher dimensional example is easily constructed in a similar way. Thus the converse of Theorem 4.2 is false; hence, exponential dichotomy is strictly stronger than exponential separation.
(ii) In the case of ordinary differential equation, the concepts of exponential dichotomy, and integral separateness are equivalent modulo the use of shifted equations (see [100]).

Finally, in the following theorem, we recall an important result in [43, Theorem 3.6] on denseness of integrally separated discrete-time linear cocycles on the set of all bounded discrete-time linear random dynamical systems.

Theorem 4.3 (Denseness of Integrally Separated Discrete-Time Linear Random Dynamical Systems) *Let* $(X, \mathcal{A}, m)$ *be a nonatomic Lebesgue probability space and* $T : X \to X$ *an ergodic transformation preserving the probability* m. *Let* $\mathcal{M} : X \to \mathrm{GL}(d)$ *be measurable matrix-valued function such that this function and its inverse are bounded. Then, for any* $\varepsilon > 0$, *there exists a measurable matrix* $\mathcal{M}_\varepsilon : X \to \mathrm{GL}(d)$ *such that*

$$\operatorname*{\mathit{ess}\,sup}_{x\in X} \|\mathcal{M}(x) - \mathcal{M}_\varepsilon(x)\| + \operatorname*{\mathit{ess}\,sup}_{x\in X} \|\mathcal{M}(x)^{-1} - \mathcal{M}_\varepsilon(x)^{-1}\| \le \varepsilon$$

and the generated discrete-time linear random dynamical system $\Phi_{\mathcal{M}_\varepsilon} : \mathbb{Z} \times X \to \mathrm{GL}(d)$ *defined by*

$$\Phi_{\mathcal{M}_\varepsilon}(n, x) = \begin{cases} \mathcal{M}_\varepsilon(T^n x) \circ \cdots \circ \mathcal{M}_\varepsilon(x), & \text{if } n > 0, \\ id, & \text{if } n = 0, \\ \mathcal{M}_\varepsilon(T^n x)^{-1} \circ \cdots \circ \mathcal{M}_\varepsilon(T^{-1}x)^{-1}, & \text{if } n < 0. \end{cases}$$

is integrally separated, i.e., there exist positive constants L, α *and an invariant decomposition*

$$\mathbb{R}^d = E_1(x) \oplus E_2(x) \oplus \cdots \oplus E_d(x)$$

such that for $i = 1, \ldots, d$, *the following assertions hold:*

(i) $\dim E_i(x) = 1$ *for* m*-a.e.* $x \in X$,
(ii) For m*-a.e.* $x \in X$, $u \in \bigoplus_{j=1}^{i} E_j(x) \setminus \{0\}$ *and* $v \in \bigoplus_{j=i+1}^{d} E_j(x) \setminus \{0\}$ *we have*

$$\|\Phi_{\mathcal{M}_\varepsilon}(n, x)u\| \ge Le^{n\alpha} \|\Phi_{\mathcal{M}_\varepsilon}(n, x)v\|, \qquad n \in \mathbb{N}.$$

4.2.2 Integral Separation for Bounded Linear Random Differential Equations

We start this subsection by extending the notion of integral separation from discrete-time bounded linear to bounded linear random differential equations; see also [1, Definition 5.3.2] and Definition 1.5 for this notion of linear nonautonomous differential equations.

Definition 4.2 (Integral Separation) System (4.1) is called *integrally separated* if there exist a set of full measure $\widehat{\Omega}$ being invariant under $(\theta_t)_{t\in\mathbb{R}}$, positive constants L, α and an invariant measurable decomposition

$$\mathbb{R}^d = W_1(\omega) \oplus W_2(\omega) \oplus \cdots \oplus W_d(\omega)$$

such that for $i = 1, \ldots, d$ the following assertions hold:

(i) $\dim W_i(\omega) = 1$ for all $\omega \in \widehat{\Omega}$,
(ii) For all $\omega \in \widehat{\Omega}$, $u \in \bigoplus_{j=1}^{i} W_j(\omega) \setminus \{0\}$ and $v \in \bigoplus_{j=i+1}^{d} W_j(\omega) \setminus \{0\}$ we have

$$\frac{\|\Phi_A(t,\omega)u\|}{\|u\|} \geq L e^{t\alpha} \frac{\|\Phi_A(t,\omega)v\|}{\|v\|}, \qquad t \in \mathbb{R}_{\geq 0}.$$

Let $\mathcal{R} \subset \mathcal{L}^\infty(\Omega, \mathbb{R}^{d\times d})$ be the set of bounded measurable matrices A such that the associated linear random differential equation $\dot{x} = A(\theta_t\omega)x$ is integrally separated. In this subsection, we focus on the question: How big is the set $\mathcal{R}$ in the complete metric space $(\mathcal{L}^\infty(\Omega, \mathbb{R}^{d\times d}), \rho_\infty)$? The answer of this question depends on the structure of the base space $(\Omega, \mathcal{F}, \mathbb{P}, (\theta_t)_{t\in\mathbb{R}})$. Before going to formulate this answer, we need the following preparation:
Firstly, in the whole section, we assume that $(\Omega, \mathcal{F}, \mathbb{P})$ is a Lebesgue probability space (or Standard probability space) and $(\theta_t)_{t\in\mathbb{R}}$ is an ergodic flow from Ω into itself without fixed points and preserving the probability $\mathbb{P}$. Hence, according to representation of ergodic flow theorem, see Theorem 3.1; there exists a Lebesgue probability space $(X, \mathcal{A}, m)$ and an ergodic transformation $T : X \to X$ preserving the probability m and a measurable function $f : X \to \mathbb{R}_{\geq 0}$ which is bounded from 0 and ∞ such that the flow $(S_t)_{t\in\mathbb{R}}$ built under a function (X, T, m, f) is isomorphic to the flow $(\theta_t)_{t\in\mathbb{R}}$, i.e., there exists a measure preserving bijective transformation $H : \Omega \to B$, where $B = \{(x, t) : 0 \leq t < f(x)\}$, such that

$$H \circ \theta_t(\omega) = S_t \circ H(\omega) \qquad \text{for all } \omega \in \Omega. \tag{4.12}$$

Since f is bounded from 0 and ∞, it follows that

$$\overline{c} := \operatorname*{ess\,sup}_{x\in X} f(x) \in (0, \infty), \quad \underline{c} := \operatorname*{ess\,inf}_{x\in X} f(x) \in (0, \infty). \tag{4.13}$$

Finally, since $(X, \mathcal{F}, m)$ is a Lebesgue probability space, $(X, \mathcal{F}, m)$ is of one of the following cases:

Case A: $(X, \mathcal{F}, m)$ is isomorphic to $([0, 1], \lambda)$, where λ is the standard Lebesgue probability on $[0, 1]$.

Case B: The set X is isomorphic to a probability space $[0, s] \cup \{x_1, x_2, \dots, x_k\}$, where $s = 1 - \sum_{i=1}^{k} p_i$ with p_i is the probability of $\{x_i\}$ and k can be equal to ∞.

Remark 4.3 Suppose that **Case B** holds. Since T is an invertible ergodic transformation on X, it follows that X has a finite atoms, and from this, we derive that X consists of only atoms. Hence, X is isomorphic to the set $\{x_1, \dots, x_k\}$ with the probability of x_i is $\frac{1}{k}$. Consequently, the base space $(\Omega, \mathcal{F}, \mathbb{P}, (\theta_t)_{t\in\mathbb{R}})$ is isomorphic to $(\mathcal{S}^1, \mathcal{B}(\mathcal{S}^1), \lambda, (R_t)_{t\in\mathbb{R}})$, where $\mathcal{S}^1 := \{e^{2\pi i z} : z \in [0, 1]\}$ is the unit circle, λ is the standard Lebesgue probability on $\mathcal{S}^1$, and $R_t : \mathcal{S}^1 \to \mathcal{S}^1$ is the rotation map defined by $R_t(e^{2\pi i z}) = e^{2\pi i(z+t)}$.

4.2.2.1 Case A: The Ergodic Flow Is Not Isomorphic to the Rotation Flow on $\mathcal{S}^1$

Theorem 4.4 *Suppose that* **Case A** *holds. Then, we have:*

(i) The set $\mathcal{R}$ is dense in $(\mathcal{L}^\infty(\Omega, \mathbb{R}^{d\times d}), \rho_\infty)$,
(ii) The set $\mathcal{R}$ is open in $(\mathcal{L}^\infty(\Omega, \mathbb{R}^{d\times d}), \rho_\infty)$.

As a consequence, the set of bounded linear random differential equations being integrally separated is generic.

Proof (i) Let $A \in \mathcal{L}^\infty(\Omega, \mathbb{R}^{d\times d})$ and $\varepsilon > 0$ be arbitrary. Our aim is to construct $A_\varepsilon \in \mathcal{L}^\infty(\Omega, \mathbb{R}^{d\times d})$ with $\rho_\infty(A, A_\varepsilon) < \varepsilon$ such that the linear differential equation $\dot{x} = A_\varepsilon(\theta_t\omega)x$ is integrally separated. For this purpose, we define a measurable matrix-valued function $\mathcal{M} : X \to \mathbb{R}^{d\times d}$ by

$$\mathcal{M}(x) := \Phi_A(f(x), H^{-1}(x, 0)) \qquad \text{for all } x \in X. \tag{4.14}$$

By Corollary 4.1, we have

$$\operatorname*{ess\,sup}_{x\in X} \|\mathcal{M}(x)\|, \operatorname*{ess\,sup}_{x\in X} \|\mathcal{M}(x)^{-1}\| \le \exp\left(\overline{c}\hat{K}\right), \tag{4.15}$$

where $\overline{c}$ is defined as in (4.13). Thus, $\mathcal{M}$ and $\mathcal{M}^{-1}$ are essentially bounded. Let δ is a positive number satisfying that

$$\delta < \frac{1}{\exp\left(\overline{c}\hat{K}\right)} \text{ and } \delta \frac{\exp(\overline{c}\hat{K})}{1 - \delta \exp(\overline{c}\hat{K})} \left(\frac{1}{\underline{c}} + 2 \operatorname*{ess\,sup}_{\omega\in\Omega} \|A(\omega)\|\right) < \varepsilon. \tag{4.16}$$

In view of Theorem 4.3, there exists a linear random mapping $\widehat{\mathcal{M}} : X \to \mathbb{R}^{d\times d}$ such that

$$\operatorname*{ess\,sup}_{x\in X} \|\mathcal{M}(x) - \widehat{\mathcal{M}}(x)\| + \operatorname*{ess\,sup}_{x\in X} \|\mathcal{M}(x)^{-1} - \widehat{\mathcal{M}}(x)^{-1}\| \le \delta, \tag{4.17}$$

and the generated linear cocyle $\Phi_{\widehat{\mathcal{M}}} : \mathbb{Z}\times X \to \mathbb{R}^{d\times d}$ over the metric dynamical system $(X, \mathcal{A}, m, T)$ is integrally separated. For any $x \in X$, we set

$$R(x,t) := \mathrm{id} - \frac{t}{f(x)}(\mathrm{id} - \widehat{\mathcal{M}}(x)\mathcal{M}(x)^{-1}), \quad t \in [0, f(x)].$$

Combining (4.15) and (4.17) yields that for m-a.e. $x \in X$

$$\|\widehat{\mathcal{M}}(x)\mathcal{M}(x)^{-1} - \mathrm{id}\| \le \delta \exp\left(\overline{c}\hat{K}\right),$$

which implies that for $t \in [0, f(x)]$

$$\|R(x,t) - \mathrm{id}\| \le \delta \exp\left(\overline{c}\hat{K}\right), \quad \left\|\frac{\partial R}{\partial t}(x,t)\right\| \le \frac{\delta}{\underline{c}} \exp\left(\overline{c}\hat{K}\right). \tag{4.18}$$

Note that $\delta \exp\left(\overline{c}\hat{K}\right) < 1$ and thus $R(t,x)$ is invertible and we have

$$\|R(x,t)^{-1}\| \le \frac{1}{1 - \delta \exp\left(\overline{c}\hat{K}\right)}, \quad \|R(x,t)^{-1} - \mathrm{id}\| \le \frac{\delta \exp\left(\overline{c}\hat{K}\right)}{1 - \delta \exp\left(\overline{c}\hat{K}\right)}. \tag{4.19}$$

We divide the remaining of the proof of this assertion into two steps:
Step 1: In this step, we construct $\widehat{A} : \Omega \to \mathbb{R}^{d\times d}$ such that $\rho_\infty(A, \widehat{A}) < \varepsilon$ and

$$\widehat{\mathcal{M}}(x) = \Phi_{\widehat{A}}(f(x), H^{-1}(x,0)) \qquad \text{for all } x \in X. \tag{4.20}$$

For this purpose, we define for all $(x,t) \in B$

$$\widehat{A}(H^{-1}(x,t)) := R(x,t)A(H^{-1}(x,t))R(x,t)^{-1} + \frac{\partial R}{\partial t}(x,t)R(x,t)^{-1}.$$

Thus,

$$\begin{aligned}\|\widehat{A}(H^{-1}(x,t)) - A(H^{-1}(x,t))\| &\le \|R(x,t) - \mathrm{id}\|\,\|A(H^{-1}(x,t))R(x,t)^{-1}\| \\ &+ \|A(H^{-1}(x,t))\|\,\|R(x,t)^{-1} - \mathrm{id}\| + \left\|\frac{\partial R}{\partial t}(x,t)R(x,t)^{-1}\right\|.\end{aligned}$$

Using (4.18) and (4.19), we obtain that

$$\operatorname*{ess\,sup}_{\omega\in\Omega}\|A(\omega)-\widehat{A}(\omega)\|\leq\delta\frac{\exp(\overline{c}\hat{K})}{1-\delta\exp(\overline{c}\hat{K})}\left(\frac{1}{\underline{c}}+2\operatorname*{ess\,sup}_{\omega\in\Omega}\|A(\omega)\|\right),$$

which together with (4.16) gives that $\rho_\infty(A,\widehat{A})<\varepsilon$. To conclude the proof of this step, let $x\in X$ be arbitrary. By definition of $\widehat{A}$ and (4.12), for any $t\in[0,f(x))$, the equation

$$\dot{\xi}=\widehat{A}(\theta_t\circ H^{-1}(x,0))\xi=\widehat{A}(H^{-1}(x,t))\xi,\quad \xi(0)=\xi_0\in\mathbb{R}^d$$

has the solution

$$\xi(t)=R(x,t)\Phi_A(t,H^{-1}(x,0))\xi_0.$$

Consequently,

$$\Phi_{\widehat{A}}(f(x),H^{-1}(x,0))=R(x,f(x))\Phi_A(f(x),H^{-1}(x,0)).$$

By definition of R, equality (4.20) holds.

Step 2: In this step, we show that the linear random dynamical system $\Phi_{\widehat{A}}(t,\omega)$ is integrally separated. From (4.12) and (4.14), we derive that for any $x\in X$ and $n\in\mathbb{N}$

$$\begin{aligned}\Phi_{\widehat{\mathcal{M}}}(n,x)&=\prod_{k=0}^{n-1}\widehat{\mathcal{M}}(T^kx)\\&=\prod_{k=0}^{n-1}\Phi_{\widehat{A}}(f(T^{k-1}x),H^{-1}(T^{k-1}x,0))\\&=\Phi_{\widehat{A}}\left(\sum_{k=0}^{n-1}f(T^kx),H^{-1}(x,0)\right).\end{aligned}$$

Hence, from the fact that $\Phi_{\widehat{\mathcal{M}}}$ is integrally separated and f is strictly separated from 0 and ∞, we can easily deduce that the linear random dynamical system $\Phi_{\widehat{A}}(t,\omega)$ is also integrally separated. The proof of (i) is complete.

(ii) Let $A\in\mathcal{R}$ be arbitrary. Define a measurable matrix-valued function $\mathcal{M}:X\to\mathbb{R}^{d\times d}$ by

$$\mathcal{M}(x):=\Phi_A(f(x),H^{-1}(x,0))\qquad\text{for all }x\in X.$$

Since the linear random dynamical system Φ_A is integrally separated and f is strictly separated from 0 and ∞, it follows that the discrete time linear random dynamical system $\Phi_{\mathcal{M}}$ generated by $\mathcal{M}$ over $(X,\mathcal{A},m,T)$ is also integrally separated. Note that for discrete time linear random dynamical systems, integral separation is a robust property; see, e.g., [44]. Thus, there exists $\varepsilon < 1$ such that for all $\widehat{\mathcal{M}}$, satisfying that

$$\operatorname*{ess\,sup}_{x\in X} \|\mathcal{M}(x) - \widehat{\mathcal{M}}(x)\| + \operatorname*{ess\,sup}_{x\in X} \|\mathcal{M}(x)^{-1} - \widehat{\mathcal{M}}(x)^{-1}\| \le \varepsilon,$$

the generated discrete time linear random dynamical system $\Phi_{\widehat{\mathcal{M}}}$ is also integrally separated. Let $\widehat{A} \in \mathcal{L}^\infty(\Omega, \mathbb{R}^{d\times d})$ be arbitrary such that $\rho_\infty(A, \widehat{A}) < \widehat{\varepsilon}$, where $\widehat{\varepsilon} < 1$ is chosen later. Analog to Lemma 4.2, there exists $\tilde{K} > 8$ (independent of $\widehat{\varepsilon}$) and a set of full measure $\widehat{\Omega}$ being invariant under $(\theta_t)_{t\in\mathbb{R}}$ such that for all $\omega \in \widehat{\Omega}$, $\beta > \alpha$, we have

$$\int_\alpha^\beta \|A(\theta_s\omega)\|\, ds, \int_\alpha^\beta \|\widehat{A}(\theta_s\omega)\|\, ds \le \tilde{K}(\beta - \alpha)$$

and

$$\int_\alpha^\beta \|A(\theta_s\omega) - \widehat{A}(\theta_s\omega)\|\, ds \le 2\widehat{\varepsilon}(\beta - \alpha).$$

Note that

$$\Phi_A(t,\omega) - \Phi_{\widehat{A}}(t,\omega) = \int_0^t A(\theta_s\omega)\Phi_A(s,\omega) - \widehat{A}(\theta_s\omega)\Phi_{\widehat{A}}(s,\omega)\, ds,$$

which implies that

$$\begin{aligned}\|\Phi_A(t,\omega) - \Phi_{\widehat{A}}(t,\omega)\| \le& \left|\int_0^t \|A(\theta_s\omega) - \widehat{A}(\theta_s\omega)\|\,\|\Phi_A(s,\omega)\|\, ds\right| \\ &+ \left|\int_0^t \|\widehat{A}(\theta_s\omega)\|\,\|\Phi_A(s,\omega) - \Phi_{\widehat{A}}(s,\omega)\|\, ds\right|.\end{aligned}$$

Hence, by Corollary 4.1, we have

$$\max_{0\le s\le |t|} \|\Phi_A(s,\omega) - \Phi_{\widehat{A}}(s,\omega)\| \le 2\widehat{\varepsilon}|t|e^{\tilde{K}|t|} + \tilde{K}|t| \max_{0\le s\le |t|} \|\Phi_A(s,\omega) - \Phi_{\widehat{A}}(s,\omega)\|.$$

Letting $|t| = \frac{1}{2\tilde{K}}$ and using $\tilde{K} > 8$ yield that

$$\max_{0\le |s|\le \frac{1}{2\tilde{K}}} \|\Phi_A(s,\omega) - \Phi_{\widehat{A}}(s,\omega)\| \le \widehat{\varepsilon} \qquad \text{for all } \omega \in \widehat{\Omega}.$$

Hence, using the cocycle property, we obtain that for m-a.e. $x \in X$

$$\begin{aligned}\|\Phi_A(f(x), H^{-1}(x,0)) - \Phi_{\widehat{A}}(f(x), H^{-1}(x,0))\| &\leq n\widehat{\varepsilon}\exp(\overline{c}\tilde{K})\\ &\leq \widehat{\varepsilon}\exp(1+3\overline{c}\tilde{K})\end{aligned}$$

and

$$\|\Phi_A(-f(x), H^{-1}(x,0)) - \Phi_{\widehat{A}}(-f(x), H^{-1}(x,0))\| \leq \widehat{\varepsilon}\exp(1+3\overline{c}\tilde{K}),$$

where $\overline{c}$ is given as in (4.13) and n is the smallest integer such that $\frac{n}{2\tilde{K}} \geq \overline{c}$. Choose $\widehat{\varepsilon} := \frac{\varepsilon}{2\exp(1+3\overline{c}\tilde{K})}$. Then for all $\widehat{A} \in \mathcal{L}^\infty(\Omega, \mathbb{R}^{d\times d})$ with $\rho_\infty(\widehat{A}, A) < \widehat{\varepsilon}$, the linear discrete-time random dynamical system generated by $\Phi_{\widehat{A}}(f(x), H^{-1}(x,0))$ is integrally separated. Analog to proof of Step 2 of (i), the linear random dynamical system $\Phi_{\widehat{A}}(t,\omega)$ is also integrally separated. The proof is complete.

4.2.2.2 Case B: The Ergodic Flow Is Isomorphic to the Rotation Flow on $\mathcal{S}^1$

Theorem 4.5 *Suppose that **Case B** holds. Then, the set $\mathcal{R}$ is open but not generic in $(\mathcal{L}^\infty(\Omega, \mathbb{R}^{d\times d}), \rho_\infty)$.*

Proof By virtue of Remark 4.3, the base space $(\Omega, \mathcal{F}, \mathbb{P}, (\theta_t)_{t\in\mathbb{R}})$ can be identified with $(\mathcal{S}^1, \mathcal{B}(\mathcal{S}^1), \lambda, (R_t)_{t\in\mathbb{R}})$. Let $A \in (\mathcal{L}^\infty(\Omega, \mathbb{R}^{d\times d}), \rho_\infty)$ be arbitrary. By periodicity of the flow $(R_t)_{t\in\mathbb{R}}$, i.e., $R_{t+1} \equiv R_t$, we have

$$\Phi_A(t, \theta_s\omega) = \Phi_A(t, \theta_{s+1}\omega) \qquad \text{for all } s \in \mathbb{R}, \omega \in \mathcal{S}^1.$$

Let $\omega_0 = e^{2\pi i 0}$ and define $M := \Phi_A(1, \omega_0)$. Thus, for any $\omega \in \mathcal{S}^1$, there exists a unique $s \in [0, 1)$ such that $\omega = \theta_s\omega_0$, and therefore we have for all $t \geq 0$

$$\Phi_A(t,\omega) = \Phi_A(t+s-n, \omega_0)M^n\Phi_A(s,\omega_0)^{-1},$$

where n is the largest integer smaller or equal to $t + s$. Therefore, the generated linear random dynamical system Φ_A is integrally separated if and only if the set of the logarithms of the moduli of eigenvalues of the matrix M has exactly d elements. Due to the fact that eigenvalues depend continuously on the matrix, the set $\mathcal{R}$ is open. Also note that the multiplicity 2 of pairs of conjugate complex eigenvalues is also stable under small perturbation. Therefore, there exists an open set in $(\mathcal{L}^\infty(\Omega, \mathbb{R}^{d\times d}), \rho_\infty)$ which has no intersection with $\mathcal{R}$. The proof is complete.

4.3 Analyticity of Generic Lyapunov Spectrum of Bounded Linear Random Dynamical Systems

4.3.1 Analyticity of the Top Lyapunov Exponent for Positive Linear Random Dynamical Systems

We start this subsection by recalling a notion of analytic function. Let X, Y be Banach spaces. For $k \in \mathbb{N}$, a function $p : X \to Y$ is called a *continuous homogeneous polynomial of degree* k if there exists a continuous k-multilinear symmetric function $\varphi : X^k \to Y$ such that $p(x) = \varphi(x, \dots, x)$.

A function $f : X \to Y$ is said to be *real analytic* at $x \in X$ if there exists $\delta > 0$ and a sequence of continuous homogeneous polynomials $(p_k)_{k\geq 0}$, where $p_k : X \to Y$ is of degree k, such that $\sum_{k=0}^{\infty} \|p_k\|\delta^k < \infty$ and

$$f(x + \Delta) = \sum_{k=0}^{\infty} p_k(\Delta) \qquad \text{for } \Delta \in X \text{ with } \|\Delta\| \leq \delta.$$

In the following lemma, we collect some fundamental properties of analytic functions which are used in the next section.

Lemma 4.4 *The following statements hold:*

- *(i) Let X, Y be Banach spaces. Then, the composition, the summation, and the subtraction of two analytic functions from X to Y are again analytic.*
- *(ii) Let X, Y be Banach spaces. A bounded linear operator $T : X \to Y$ is analytic.*
- *(iii) Let $f, g : \mathcal{L}^\infty(\Omega, \mathbb{R}^{d\times d}) \to \mathcal{L}^\infty(\Omega, \mathbb{R}^{d\times d})$ be analytic functions. Then, the product function $fg : \mathcal{L}^\infty(\Omega, \mathbb{R}^{d\times d}) \to \mathcal{L}^\infty(\Omega, \mathbb{R}^{d\times d})$, $A \mapsto fg(A)$ defined by $fg(A)(\omega) := f(A)(\omega)g(A)(\omega)$ is also analytic.*

Proof We can find a proof of (i) in, e.g., [125]. Meanwhile, the statements (ii) and (iii) follow directly from the definition of analytic function.

In the rest of this subsection, we recall a version of *Perron-Frobenious theorem* about analyticity of the top Lyapunov exponent of positive linear random dynamical systems. For this purpose, a subset $C \subset \mathbb{R}^d$ is called a *proper cone* if $tv \in C$ for all $t \geq 0$, $v \in C$ and $C \cap (-C) = \{0\}$. For a closed proper convex cone C, the dual cone C' is defined by

$$C' := \{x \in \mathbb{R}^d : \langle x, v\rangle \geq 0 \quad \text{for all } v \in C\}.$$

Recall that the *Hilbert metric* d_C on $C \setminus \{0\}$ is defined by

$$d_C(x, y) := \log(\beta_C(x, y)) + \log(\beta_C(y, x)) \qquad \text{for } x, y \in C \setminus \{0\},$$

where $\beta_C(x, y) := \inf\{t > 0 : tx - y \in C\}$. In what follows, the Hilbert metric d_C will be understood in a wider sense that $d_C(\pm x, \pm y) := d_C(x, y)$ for all $(x, y) \in C$.

Theorem 4.6 (Perron-Frobenious Theorem for Positive Linear Random Dynamical Systems) *Let $(C_x)_{x\in X}$ be a family of closed proper convex cones satisfying the following conditions:*

(C1) there is a measurable mapping $c : X \to \mathbb{R}^d$ such that $\|c(x)\| = 1$ and $B(c(x), r)$ $\subset C_x$ with r is independent of x,

(C2) there is a measurable mapping $c' : X \to \mathbb{R}^d$ such that $\|c'(x)\| = 1$ and $B(c'(x), r') \subset C'_x$ with r' is independent of x.

Let $A \in L^\infty(X, \mathrm{GL}(d))$ satisfy that $A(x)C_x \subset C_{Tx} \cup (-C_{Tx})$ and there exists $R < \infty$ such that

$$d_{C_{Tx}}(A(x)v, c(Tx)) \leq R \qquad \text{for all } x \in X, v \in C_x \setminus \{0\}.$$

Then, the following statements hold:

(i) The Oseledets subspace corresponding to the top Lyapunov exponent $\lambda_1(A)$ is of dimensional one,

(ii) The map $\lambda_1(\cdot)$ is analytic at A.

Proof For each $x \in X$, let

$$r(x) := \begin{cases} 1, & \text{if } A(x)C_x \subset C_{Tx}, \\ -1, & \text{if } A(x)C_x \subset -C_{Tx}. \end{cases}$$

By properties of measurable functions c and c' in (C1) and (C2), we have

$$r(x) = \begin{cases} 1, & \text{if } \langle A(x)c(x), c'(Tx)\rangle > 0, \\ -1, & \text{if } \langle A(x)c(x), c'(Tx)\rangle < 0, \end{cases}$$

which implies that r is measurable. Now, we consider a linear operator $\mathcal{T} : \mathcal{L}^\infty(X, \mathbb{R}^{d\times d}) \to \mathcal{L}^\infty(X, \mathbb{R}^{d\times d})$ defined by

$$\mathcal{T}M(x) = r(x)M(x) \quad \text{for all } M \in \mathcal{L}^\infty(X, \mathbb{R}^{d\times d}).$$

Obviously, $\lambda_1(\mathcal{T}M) = \lambda_1(M)$ and the Oseledets subspaces of M and $\mathcal{T}M$ coincide. Hence, it is sufficient to prove (i) and (ii) for $\mathcal{T}A$. By definition of r, the random matrix $\mathcal{T}A$ preserves the family of cones C_x in the sense that $\mathcal{T}A(x)C_x \subset C_{Tx}$. Hence, using results in [10] yields (i). Meanwhile, the assertion (ii) is obtained by applying the result in [60]. The proof is complete.

4.3.2 Analyticity of Lyapunov Exponents for Discrete-Time Bounded Linear Random Dynamical Systems

This subsection is devoted to studying discrete-time bounded linear random dynamical systems introduced in Sect. 4.1.1. The main result in this subsection is to show that generically Lyapunov exponents of discrete-time bounded linear random dynamical systems depend analytically on the generators. To achieve this result, we show that in the set of integrally separated bounded linear random dynamical systems, the Lyapunov exponents depend analytically on the generator. This assertion together with the result on the genericity of integral separation in the space of bounded invertible random matrices (see Theorem 4.3 and [43]) and Lemma 4.1 implies that analyticity of Lyapunov exponents is a generic property.

In what follows, we introduce and prove some fundamental properties of a specific cone in the exterior power space $\Lambda^k\mathbb{R}^d$. This cone is a key object in the proof of the main result in which we show that the generated random dynamical system of any integrally separated bounded linear random dynamical system on $\Lambda^k\mathbb{R}^d$ after a long enough iteration and a random change of coordinate preserves this cone.

Lemma 4.5 *For each $k \in \{1, \dots, d\}$, we define a subset $C_k \subset \Lambda^k\mathbb{R}^d$ by*

$$C_k := \left\{ \sum_{1\le i_1<\cdots<i_k\le d} : \alpha_{i_1\dots i_k} e_{i_1} \wedge \cdots \wedge e_{i_k} : \alpha_{1\dots k} = \max_{1\le i_1<\cdots<i_k\le d} |\alpha_{i_1\dots i_k}| \right\}. \tag{4.21}$$

Then, the following statements hold:

(i) C_k is a closed proper convex cone.
(ii) The dual cone C_k' of C_k is given by

$$C_k' := \left\{ \sum_{1\le i_1<\cdots<i_k\le d} : \alpha_{i_1\dots i_k} e_{i_1} \wedge \cdots \wedge e_{i_k} : \alpha_{1\dots k} \ge \frac{1}{2} \sum_{1\le i_1<\cdots<i_k\le d} |\alpha_{i_1\dots i_k}| \right\}.$$

Consequently, $e_1 \wedge \cdots \wedge e_k$ is an interior point of both C_k and C_k'. More, precisely, we have

$$B\left(e_1 \wedge \cdots \wedge e_k, \frac{1}{3}\right) \subset C_k \quad and \quad B(e_1 \wedge \cdots \wedge e_k, \frac{1}{3C_d^k}) \subset C_k', \tag{4.22}$$

where $C_d^k := \frac{d!}{k!(d-k)!}$.

Proof The assertion (i) follows directly from the definition of C_k and the space $\Lambda^k\mathbb{R}^d$; see Sect. 3.2.1. To prove (ii), we compute the dual cone C_k' of C_k. For this purpose, let $v = \sum_{1\le i_1<\cdots<i_k\le d} \alpha_{i_1\dots i_k} e_{i_1} \wedge \cdots \wedge e_{i_k} \in C_k'$ be arbitrary. Let $x = \sum_{1\le i_1<\cdots<i_k\le d} \beta_{i_1\dots i_k} e_{i_1} \wedge \cdots \wedge e_{i_k}$, where $\beta_{1\dots k} = 1$ and for $1 \le i_1 < \cdots < i_k \le d$

with $(i_1, \dots, i_k) \neq (1, \dots k)$

$$\beta_{i_1 \dots i_k} = \begin{cases} 1 & \text{if } \alpha_{i1 \dots i_k} \geq 0, \\ -1 & \text{if } \alpha_{i1 \dots i_k} < 0. \end{cases}$$

Obviously, $x \in C_k$ and a direct computation yield that

$$\langle x, v \rangle = \alpha_{1 \dots k} + \sum_{1 \leq i_1 < \dots < i_k \leq d, (i_1, \dots, i_k) \neq (1, \dots, k)} |\alpha_{i_1 \dots i_k}|.$$

Since $\langle x, v \rangle \geq 0$, it follows that $\alpha_{1 \dots k} \geq \frac{1}{2} \sum_{1 \leq i_1 < \dots < i_k \leq d} |\alpha_{i_1 \dots i_k}|$. Conversely, let $v = \sum_{1 \leq i_1 < \dots < i_k \leq d} \alpha_{i_1 \dots i_k} e_{i_1} \wedge \dots \wedge e_{i_k}$ satisfy that $\alpha_{1 \dots k} \geq \frac{1}{2} \sum_{1 \leq i_1 < \dots < i_k \leq d} |\alpha_{i_1 \dots i_k}|$. For any $x = \sum_{1 \leq i_1 < \dots < i_k \leq d} \beta_{i_1 \dots i_k} e_{i_1} \wedge \dots \wedge e_{i_k} \in C_k$, we have

$$\begin{aligned} \langle x, v \rangle &= \sum_{1 \leq i_1 < \dots < i_k \leq d} \alpha_{i_1 \dots i_k} \beta_{i_1 \dots i_k} \\ &\geq \alpha_{1 \dots k} \beta_{1 \dots k} - \sum_{1 \leq i_1 < \dots < i_k \leq d, (i_1, \dots, i_k) \neq (1, \dots k)} |\alpha_{i_1 \dots i_k} \beta_{i_1 \dots i_k}| \\ &\geq \beta_{1 \dots k} \left(\alpha_{1 \dots k} - \sum_{1 \leq i_1 < \dots < i_k \leq d, (i_1, \dots, i_k) \neq (1, \dots k)} |\alpha_{i_1 \dots i_k}| \right), \end{aligned}$$

where we use the fact that $\beta_{1 \dots k} = \max_{1 \leq i_1 < \dots < i_k \leq d} |\beta_{i_1 \dots i_k}|$ to obtain the preceding inequality. Therefore, $\langle x, v \rangle \geq 0$ and the assertion (ii) is verified. In the remaining part of the proof, we show (4.22). For this purpose, let $v = \sum_{1 \leq i_1 < \dots < i_k \leq d} \alpha_{i_1 \dots i_k} e_{i_1} \wedge \dots \wedge e_{i_k}$ satisfy that $\|e_1 \wedge \dots \wedge e_k - v\|_{\Lambda^k \mathbb{R}^d} \leq \frac{1}{3}$. Thus, from

$$\|e_1 \wedge \dots \wedge e_k - v\|^2_{\Lambda^k \mathbb{R}^d} = (\alpha_{1 \dots k} - 1)^2 + \sum_{1 \leq i_1 < \dots < i_k \leq d, (i_1, \dots, i_k) \neq (1, \dots k)} \alpha^2_{i_1 \dots i_k}$$

we derive that

$$\frac{2}{3} \leq \alpha_{1 \dots k} \quad \text{and} \quad \max_{1 \leq i_1 < \dots < i_k \leq d, (i_1, \dots, i_k) \neq (1, \dots, k)} |\alpha_{i_1 \dots i_k}| \leq \frac{1}{3}. \tag{4.23}$$

Consequently, $v \in C_k$ and thus $B(e_1 \wedge \dots \wedge e_k, \frac{1}{3}) \subset C_k$. To prove the remaining assertion in (4.22), let $v = \sum_{1 \leq i_1 < \dots < i_k \leq d} \alpha_{i_1 \dots i_k} e_{i_1} \wedge \dots \wedge e_{i_k}$ satisfy that $\|e_1 \wedge \dots \wedge e_k - v\|_{\Lambda^k \mathbb{R}^d} \leq \frac{1}{3C_d^k}$. Analog to the arguments yielding (4.23), we have

$$1 - \frac{1}{3C_d^k} \leq \alpha_{1\ldots k} \quad \text{and} \quad \max_{1\leq i_1<\cdots<i_k\leq d,(i_1,\ldots,i_k)\neq(1,\ldots,k)} |\alpha_{i_1\ldots i_k}| \leq \frac{1}{3C_d^k},$$

which implies that $\alpha_{1\ldots k} \geq \sum_{1\leq i_1<\cdots<i_k\leq d,(i_1,\ldots,i_k)\neq(1,\ldots,k)} |\alpha_{i_1\ldots i_k}|$. Thus, $v \in C'_k$ and the proof is complete.

Theorem 4.7 (Analyticity for Lyapunov Exponents of Generic Discrete-Time Linear Random Dynamical Systems) *Let $A \in \mathcal{L}^\infty(X, \mathrm{GL}(d))$ be an integrally separated bounded linear random dynamical system. Then, for $i = 1, \ldots, d$, the map $\gamma_i(\cdot)$ is analytic at A. As a consequence, analyticity of Lyapunov exponents is a generic property in the Banach space $(\mathcal{L}^\infty(X, \mathbb{R}^{d\times d}), \|\cdot\|_\infty)$.*

Proof Since A is integrally separated, there exist $K, \alpha > 0$ and an invariant measurable decomposition $\mathbb{R}^d = \oplus_{i=1}^d E_i(x)$, where $E_i(x)$ is a linear subspace of dimension 1, such that for all $n \in \mathbb{N}$ and m-a.e. $x \in X$ the following inequality

$$\frac{\|\Phi_A(n,x)u\|}{\|u\|} \geq Ke^{\alpha n}\frac{\|\Phi_A(n,x)v\|}{\|v\|} \tag{4.24}$$

holds for $u \in \oplus_{j=1}^i E_j(x)\setminus\{0\}$, $v \in \oplus_{j=i+1}^d E_j(x)\setminus\{0\}$, where $i = 1, \ldots, d-1$. For $i = 1, \ldots, d$, choose and fix a measurable unit vector $u_i(x) \in E_i(x)$. By invariance of $E_i(x)$, we have

$$A(x)u_i(x) = a_i(x)u_i(Tx) \qquad \text{for } i = 1, \ldots, d \text{ and } x \in X,$$

where $a_i : X \to \mathbb{R}$ is measurable. Since $A \in \mathcal{L}^\infty(X, \mathrm{GL}(d))$, there exists $M_1 > 0$ such that

$$|a_i(x)|, \frac{1}{|a_i(x)|} \leq M_1 \qquad \text{for } i = 1, \ldots, d \text{ and } x \in X \tag{4.25}$$

and by (4.24)

$$\frac{|a_i(T^{n-1}x)\ldots a_i(x)|}{|a_{i+1}(T^{n-1}x)\ldots a_{i+1}(x)|} \geq Ke^{\alpha n} \quad \text{for } i = 1, \ldots, d-1, n \in \mathbb{N}, x \in X. \tag{4.26}$$

Define $L : X \to \mathrm{GL}(d)$ by

$$L(x)e_i := u_i(x) \qquad \text{for } i = 1, \ldots, d \text{ and } x \in X,$$

where $e_1, \ldots, e_d$ is the standard Euclidean orthogonal basic of $\mathbb{R}^d$. From boundedness of A, A^{-1} and (4.24), for $i = 1, \ldots, d$, the angle between $E_i(x)$ and $\oplus_{j=1,j\neq i}^d E_j(x)$ is uniformly bounded away from zero; see Proposition 4.1. Consequently, there exists $M_2 > 0$ such that

$$\|L(x)\|, \|L^{-1}(x)\| \leq M_2 \qquad \text{for } x \in X. \tag{4.27}$$

Define a linear operator $Q_L : \mathcal{L}^\infty(X, \mathbb{R}^{d\times d}) \to \mathcal{L}^\infty(X, \mathbb{R}^{d\times d})$ by

$$Q_L M(x) := L^{-1}(Tx)M(x)L(x).$$

Then, Q_L is invertible, and from (4.27), we have $\|Q_L\|_\infty, \|Q_L^{-1}\|_\infty \le M_2^2$. Therefore, to complete the proof, it is sufficient to show that for $i = 1, \dots, d$, the map $\gamma_i(\cdot)$ is analytic at $D \in \mathcal{L}^\infty(X, \mathrm{GL}(d))$ which is defined by

$$D(x) := L^{-1}(Tx)A(x)L(x) = \mathrm{diag}(a_1(x), \dots, a_d(x)).$$

The proof of this fact is divided into two steps:
Step 1. Choose and fix an arbitrary $k \in \{1, \dots, d\}$. Let N be a positive integer such that $Ke^{\alpha N} \ge 2$. Define a random map $D^{(N)} \in \mathcal{L}^\infty(X, \mathrm{GL}(d))$ by

$$D^{(N)}(x) := D(T^{N-1}x)\dots D(x). \tag{4.28}$$

Let $\Lambda^k D^{(N)}$ denote the generated random map of $D^{(N)}$ in $\mathcal{L}^\infty(X, \mathcal{L}(\Lambda^k\mathbb{R}^d))$. In this step, we show that $\Lambda^k D^{(N)}$ preserves the cone C_k defined as in (4.21), i.e.,

$$\Lambda^k D^{(N)}(x)C_k \subset \big(C_k \cup (-C_k)\big) \quad \text{for } x \in X \tag{4.29}$$

and we also show that there exists $R > 0$ such that

$$d_{C_k}\big(\Lambda^k D^{(N)}(x)v, e_1 \wedge \cdots \wedge e_k\big) < R \text{ for } x \in X, v \in C_k \setminus \{0\} \text{ with } \|v\|_{\Lambda^k\mathbb{R}^d} = 1. \tag{4.30}$$

Choose and fix an arbitrary $x \in X$. By linearity of $D^{(N)}(x)$, to prove (4.29), it is sufficient to show that

$$\Lambda^k D^{(N)}(x)v \in \big(C_k \cup (-C_k)\big) \quad \text{for } v \in C_k \setminus \{0\} \text{ with } \|v\|_{\Lambda^k\mathbb{R}^d} = 1.$$

Let $v = \sum_{1\le i_1<\cdots<i_k\le d} : \alpha_{i_1\dots i_k} e_{i_1} \wedge \cdots \wedge e_{i_k} \in C_k \setminus \{0\}$ satisfy that $\|v\|_{\Lambda^k\mathbb{R}^d} = 1$. By definition of $\|\cdot\|_{\Lambda^k\mathbb{R}^d}$, see Sect. 3.2.1; we have

$$\|v\|^2_{\Lambda^k\mathbb{R}^d} = \sum_{1\le i_1<\cdots<i_k\le d} \alpha^2_{i_1\dots i_k} = 1,$$

which together with the fact that $\alpha_{1\dots k} = \max_{1\le i_1<\cdots<i_k\le d} |\alpha_{i_1\dots i_k}|$ implies that $\frac{1}{\sqrt{C_d^k}} \le \alpha_{1\dots k} \le 1$, where $C_d^k := \frac{d!}{k!(d-k)!}$. By definition of $D^{(N)}$ and the generated random dynamical system $\Lambda^k D^{(N)}$ on $\Lambda^k\mathbb{R}^d$, see Sect. 3.2.1; we have

$$\Lambda^k D^{(N)}(x)v = \sum_{1\le i_1<\cdots<i_k\le d} \alpha_{i_1\dots i_k} \Lambda^k D^{(N)}(x)e_{i_1} \wedge \cdots \wedge e_{i_k}$$

$$= \sum_{1\le i_1<\cdots<i_k\le d} \left(\alpha_{i_1\ldots i_k} \prod_{j=1}^{k} a_{i_j}(T^{N-1}x)\ldots a_{i_j}(x)\right) e_{i_1}\wedge\cdots\wedge e_{i_k}.$$

Using (4.26), we obtain that

$$\prod_{j=1}^{k} |a_j(T^{N-1}x)\ldots a_j(x)| \ge K e^{\alpha N} \prod_{j=1}^{k} |a_{i_j}(T^{N-1}x)\ldots a_{i_j}(x)|$$

for any index set $(i_1,\ldots,i_k) \neq (1,\ldots,k)$. This together with the fact that $\alpha_{1\ldots k} = \max_{1\le i_1<\cdots<i_k\le d} |\alpha_{i_1\ldots i_k}|$ and $Ke^{\alpha N} \ge 2$ implies that for any index set $(i_1,\ldots,i_k) \neq (1,\ldots,k)$

$$\alpha_{1\ldots k} \prod_{j=1}^{k} |a_j(T^{N-1}x)\ldots a_j(x)| \ge 2|\alpha_{i_1\ldots i_k}| \prod_{j=1}^{k} |a_{i_j}(T^{N-1}x)\ldots a_{i_j}(x)|. \tag{4.31}$$

Consequently, $\Lambda^k D^{(N)}(x)v \in \big(C_k \cup (-C_k)\big)$ and (4.29) are proved. Furthermore, from (4.31) and a direct computation yields that

$$2\alpha_{1\ldots k} \prod_{j=1}^{k} |a_j(T^{N-1}x)\ldots a_j(x)| e_1\wedge\cdots\wedge e_k - \Lambda^k D^{(N)}(x)v \in C_k,$$

$$\frac{2}{\alpha_{1\ldots k}\prod_{j=1}^{k} |a_j(T^{N-1}x)\ldots a_j(x)|}\Lambda^k D^{(N)}(x)v - e_1\wedge\cdots\wedge e_k \in C_k.$$

Thus, the distance (with respect to the Hilbert metric on the cone C_k) between two vectors $\Lambda^k D^{(N)}(x)v$ and $e_1\wedge\cdots\wedge e_k$ can be estimated as follows

$$d_{C_k}(e_1\wedge\cdots\wedge e_k, \Lambda^k D^{(N)}(x)v) \le \log\left(2\alpha_{1\ldots k}\prod_{j=1}^{k} |a_j(T^{N-1}x)\ldots a_j(x)|\right) + \log\left(\frac{2}{\alpha_{1\ldots k}\prod_{j=1}^{k} |a_j(T^{N-1}x)\ldots a_j(x)|}\right)$$

This together with (4.25) and the fact that $\frac{1}{\sqrt{C_d^k}} \le \alpha_{1\ldots k} \le 1$ shows (4.30) with

$$R := \log(2M_1^k) + \log(2\sqrt{C_d^k} M_1^k).$$

Step 2. So far, we have proved that for any $k \in \{1, \dots, d\}$, the random linear map $\Lambda^k D^{(N)}$ preserves the cone C_k. On the other hand, by Lemma 4.5, the cone C_k satisfies conditions (C1) and (C2) of Theorem 4.6 with the measurable mappings $c : X \to C_k$ and $c' : X \to C_k'$ defined by

$$c(x) = c'(x) := e_1 \wedge \cdots \wedge e_k \quad \text{for } x \in X.$$

To apply Theorem 4.6 to the random map $\Lambda^k D^{(N)}$ over $(X, \mathcal{A}, m, T^N)$, we only need to check the ergodicity of the transformation T^N. In what follows, we consider two separated cases, and our aim is to show that the map $\gamma_1(\cdot)+\cdots+\gamma_k(\cdot)$ is analytic at D, where $k \in \{1, \dots, d\}$.

Case 1. T^N is an ergodic transformation from X into itself. In this case, using the multiplicative ergodic theorem, the Lyapunov exponents of the random dynamical system generated by $M \in \mathcal{L}^\infty(X, \mathbb{R}^{d\times d})$ over $(X, \mathcal{A}, m, T^N)$, denoted by $\gamma_d^{(N)}(M) \le \cdots \le \gamma_1^{(N)}(M)$, are well-defined. Define a map $\pi : \mathcal{L}^\infty(X, \mathbb{R}^{d\times d}) \to \mathcal{L}^\infty(X, \mathbb{R}^{d\times d})$ by

$$\pi(M)(x) := M(T^{N-1}x)\dots M(x).$$

Note that π is a product of bounded linear operators

$$\pi_i : \mathcal{L}^\infty(X, \mathbb{R}^{d\times d}) \to \mathcal{L}^\infty(X, \mathbb{R}^{d\times d}), \quad M \mapsto \pi_i(M)$$

with $\pi_i(M)(x) := M(T^i x)$ and $i = 1, \dots, N-1$. Thus, by Lemma 4.4 (ii) and (iii), the function π is analytic. Furthermore, by definition of $\gamma_i^{(N)}$, we have

$$\gamma_i(M) = \frac{1}{N}\gamma_i^{(N)}(\pi(M)) \quad \text{for } i = 1, \dots, d \text{ and } M \in \mathcal{L}^\infty(X, \mathbb{R}^{d\times d}).$$

In view of Theorem 3.3, we have for all $\Delta \in \mathcal{L}^\infty(X, \mathbb{R}^{d\times d})$

$$\sum_{i=1}^k \gamma_i(D+\Delta) = \frac{1}{N}\sum_{i=1}^k \gamma_i^{(N)}(\pi(D+\Delta)) = \frac{1}{N}\gamma_1^{(N)}(\Lambda^k \pi(D+\Delta)). \tag{4.32}$$

By (4.28), we have $\pi(D) = D^{(N)}$. Using achievements proved in **Step 1** and Theorem 4.6, the map $\gamma_1^{(N)}$ is analytic in a neighborhood of $\Lambda^k\pi(D)$. Hence, by Lemma 4.4 and (4.32), the map $\sum_{i=1}^k \gamma_i(\cdot)$ is analytic at D, so the analyticity of functions $\gamma_i(\cdot)$, $i = 1, \dots, d$, at D is proved.

Case 2. T^N is not ergodic. Since all statements above still work if N is replaced by a larger integer number, hence, we can assume additionally that N is a prime number. Let W be a measurable set satisfying that $m(W) \in (0, 1)$ and $W = T^N W$. We prove by induction that for $k = 1, \dots, N$, there exists a measurable set W_k such that

$$T^N W_k = W_k \quad \text{and} \quad m(W_k) \in (0,1), m\big(W_k \cap \bigcup_{i=1}^{k-1} T^i W_k\big) = 0. \tag{4.33}$$

Obviously, (4.33) holds for $k = 1$ with $W_1 := W$ and suppose that it holds for some $k \in \{1, \ldots, N-1\}$ with a measurable set W_k. By ergodicity of T and the fact that $T\big(\bigcap_{i=1}^{N-1} T^i(W_k)\big) = \bigcap_{i=1}^{N-1} T^{i+1}(W_k) = \bigcap_{i=1}^{N-1} T^i(W_k)$, we have $m(\bigcap_{i=1}^{N-1} T^i(W_k))$
$= 0$. Since N is a prime number and $T^N W_k = W_k$, it follows that

$$\bigcap_{i=1}^{N-1} T^i(W_k) = \bigcap_{i=1}^{N-1} T^{ik}(W_k).$$

Then, let $\ell \in \{1, \ldots, N-1\}$ be the smallest integer such that $m\big(W_k \cap \bigcap_{i=1}^{\ell} T^{ik} W_k\big) = 0$. Hence, a measurable set $W_{k+1} := W_k \cap \bigcap_{i=1}^{\ell-1} T^{ik} W_k$ is also invariant under T^N and satisfies (4.33) for $k+1$. For the measurable set $\widehat{W} := W_N$, we have $T^N \widehat{W} = \widehat{W}$ and the sets $\widehat{W}, \ldots, T^{N-1}\widehat{W}$ are pairwise disjoint. From $T(\bigcup_{i=1}^{N-1} T^i \widehat{W}) = \bigcup_{i=1}^{N-1} T^i \widehat{W}$ and ergodicity of T, we have $m\big(\bigcup_{i=1}^{N-1} T^i \widehat{W}\big) = 1$. On the other hand, for any subset $U \subset \widehat{W}$ with positive measure and $m(\widehat{W} \setminus U) > 0$, we have $m\big(\bigcup_{i=1}^{N-1} T^i U\big) < 1$. This implies together with ergodicity of T that

$$m\big(\bigcup_{i=1}^{N-1} T^i U \Delta \bigcup_{i=2}^{N} T^i U\big) > 0.$$

Consequently, $m(T^N U \Delta U) \neq 0$ and therefore the restriction of the map T^N on the set $\widehat{W}$ is ergodic. Thus, using a similar statement as in *Case 1* to the induced random dynamical systems on $(\widehat{W}, \mathcal{A}, m, T^N)$ completes the proof.

4.3.3 Analyticity of Lyapunov Exponents for Bounded Linear Random Differential Equations

Our aim in this subsection is to study the analyticity of Lyapunov exponents of bounded linear random differential equations over a metric dynamical system $(\Omega, \mathcal{F}, \mathbb{P}, (\theta_t)_{t\in\mathbb{R}})$. Similarly to the setting in Sect. 4.2.2, by using the Representation of Ergodic Flow Theorem (see Theorem 3.1), there exists a Lebesgue probability space $(X, \mathcal{A}, m)$ and an ergodic transformation $T : X \to X$ preserving the probability m and a measurable function $f : X \to \mathbb{R}_{\geq 0}$ which is bounded from 0 and ∞ such that the flow $(S_t)_{t\in\mathbb{R}}$ built under a function (X, T, m, f) is isomorphic to the flow $(\theta_t)_{t\in\mathbb{R}}$, i.e., there exists a measure preserving bijective transformation $H : \Omega \to B$, where $B = \{(x, s) : 0 \leq s < f(x)\}$, such that

$$H \circ \theta_t(\omega) = S_t \circ H(\omega) \qquad \text{for all } \omega \in \Omega. \tag{4.34}$$

Since f is bounded from 0 and ∞, it follows that

$$\overline{c} := \operatorname*{ess\,sup}_{x \in X} f(x) \in (0, \infty), \quad \underline{c} := \operatorname*{ess\,inf}_{x \in X} f(x) \in (0, \infty). \tag{4.35}$$

Finally, since $(X, \mathcal{A}, m)$ is a Lebesgue probability space, $(X, \mathcal{F}, m)$ is of one of the following cases:

Case A. $(X, \mathcal{A}, m)$ is isomorphic to $([0, 1], \lambda)$, where λ is the standard Lebesgue probability on $[0, 1]$.

Case B. $(X, \mathcal{A}, m)$ is isomorphic to a probability space $[0, s] \cup \{x_1, x_2, \dots, x_k\}$, where $s = 1 - \sum_{i=1}^k p_i$ with p_i is the probability of $\{x_i\}$ and k can be equal to ∞. Note that in this case, the base space $(\Omega, \mathcal{F}, \mathbb{P}, (\theta_t)_{t \in \mathbb{R}})$ is isomorphic to $(\mathcal{S}^1, \mathcal{B}(\mathcal{S}^1), \lambda, (R_t)_{t \in \mathbb{R}})$, where $\mathcal{S}^1 := \{e^{2\pi i z} : z \in [0, 1]\}$ is the unit circle, λ is the standard Lebesgue probability on $\mathcal{S}^1$, and $R_t : \mathcal{S}^1 \to \mathcal{S}^1$ is the rotation map defined by $R_t(e^{2\pi i z}) = e^{2\pi i (z+t)}$; see Remark 4.3.

Theorem 4.8 (Analyticity for Lyapunov Exponents of Generic Bounded Linear Random Differential Equations) *For each $A \in \mathcal{L}^\infty(\Omega, \mathbb{R}^{d \times d})$, let $\gamma_d(A) \le \cdots \le \gamma_1(A)$ denote the Lyapunov exponents of the linear random differential equation*

$$\dot{\xi} = A(\theta_t \omega)\xi. \tag{4.36}$$

Then, the following statements hold:

*(i) Suppose that **Case A** holds. Then, there exists an open and dense set $\mathcal{R} \subset \mathcal{L}^\infty(\Omega, \mathbb{R}^{d \times d})$ such that the functions $\gamma_1(\cdot), \dots, \gamma_d(\cdot)$ are analytic at $A \in \mathcal{R}$.*

*(ii) Suppose that **Case B** holds. Then, the functions $\gamma_1(\cdot), \dots, \gamma_d(\cdot)$ are analytic at all $A \in \mathcal{L}^\infty(\Omega, \mathbb{R}^{d \times d})$.*

Proof Define a function $\mathcal{T} : \mathcal{L}^\infty(\Omega, \mathbb{R}^{d \times d}) \to \mathcal{L}^\infty(X, \mathbb{R}^{d \times d})$ by

$$\mathcal{T}(A)(x) := \Psi_A(f(x), H^{-1}(x, 0)) \qquad \text{for all } x \in X,$$

where Φ_A denote the linear random dynamical system generated by (4.36) and H is defined as in (4.34). By variation of constants formula, we have for all $\Delta \in \mathcal{L}^\infty(\Omega, \mathbb{R}^{d \times d})$

$$\begin{aligned}\mathcal{T}(A + \Delta)(x) &= \Psi_{A+\Delta}(f(x), H^{-1}(x, 0)) \\ &= \Psi_A(f(x), H^{-1}(x, 0)) + \\ \int_0^{f(x)} &\Psi_A(f(x) - s, \theta_s H^{-1}(x, 0))\Delta(\theta_s H^{-1}(x, 0))\Psi_A(s, H^{-1}(x, 0))\, ds\end{aligned}$$

$$= \mathcal{T}(A)(x) + \mathcal{L}_A(\Delta)(x),$$

where the operator $\mathcal{L}_A : \mathcal{L}^\infty(\Omega, \mathbb{R}^{d\times d}) \to \mathcal{L}^\infty(X, \mathbb{R}^{d\times d})$ is defined by

$$\Delta \mapsto \int_0^{f(x)} \Psi_A(f(x)-s, \theta_s H^{-1}(x,0))\Delta(\theta_s H^{-1}(x,0))\Psi_A(s, H^{-1}(x,0))\, ds.$$

Obviously, $\mathcal{L}_A$ is a linear operator, and from the boundedness of A and (4.35), the operator $\mathcal{L}_A$ is bounded. Consequently, by Lemma 4.4 (ii), the function $\mathcal{T}$ is analytic. In the remaining part of the proof, we consider two separated cases:

(i) Suppose that **Case A** holds. As is proved in [45, Theorem 3.2], there exists an open and dense set $\mathcal{R} \subset \mathcal{L}^\infty(\Omega, \mathbb{R}^{d\times d})$ such that for all $A \in \mathcal{R}$, the generated random dynamical system Ψ_A and hence also $\mathcal{T}(A)$ are integrally separated. This together with analyticity of $\mathcal{T}$, Theorem 4.7 and Lemma 4.4, completes the proof of this part.

(ii) Suppose that **Case B** holds. In this case, $(\Omega, \mathcal{F}, \mathbb{P}, (\theta_t)_{t\in\mathbb{R}})$ can be identified with $(\mathcal{S}^1, \mathcal{B}(\mathcal{S}^1), \lambda, (R_t)_{t\in\mathbb{R}})$. Choose and fix an arbitrary $\omega_0 \in \mathcal{S}^1$, and consider a map $\mathcal{P} : \mathcal{L}^\infty(\Omega, \mathbb{R}^{d\times d}) \to \mathbb{R}^{d\times d}$ defined by $\mathcal{P}A := \Psi_A(1, \omega_0)$. This is clear that the Lyapunov exponents of the linear random dynamical systems generated by

$$\dot{\xi} = A(\theta_t\omega)\xi$$

coincide with the set of the modulus of the eigenvalues of $\mathcal{P}A$. Analog to the proof of the analyticity of the map $\mathcal{T}$ above, the map $\mathcal{P}$ is also analytic. This together with the fact that the eigenvalues of matrices depend analytically on the matrices (see, e.g., [79]) proves (ii). The proof is complete.

Chapter 5
Pitchfork and Hopf Bifurcation Under Additive Noise

Abstract In this chapter, we study bifurcation of a model of pitchfork bifurcation with additive noise

$$\mathrm{d}x = \big(\alpha x - x^3\big)\mathrm{d}t + \sigma \mathrm{d}W_t\,,$$

and a model of Hopf bifurcation with additive noise of the form

$$\begin{aligned}\mathrm{d}x &= (\alpha x - \beta y - (ax - by)(x^2 + y^2))\,\mathrm{d}t + \sigma\,\mathrm{d}W_t^1\,,\\ \mathrm{d}y &= (\alpha y + \beta x - (bx + ay)(x^2 + y^2))\,\mathrm{d}t + \sigma\,\mathrm{d}W_t^2\,,\end{aligned}$$

near the underlying deterministic bifurcation point $\alpha = 0$. The bifurcation is characterized by a loss of uniform attractivity of the global random attractor, a change of experimental observability of the Lyapunov exponent, and a change of the dichotomy spectrum of the linearization along the global random attractor. The materials of this part are taken from Callaway et al. (Annales de l'Institut Henri Poincaré, Probabilités et Statistiques 53:1548–1574, 2017), Doan et al. (Nonlinearity 31:4567–4601, 2018).

5.1 Generation of a Random Dynamical System with a Random Attractor

This section is devoted to showing that both the model of pitchfork bifurcation and Hopf bifurcation with additive noise generate one-sided random dynamical systems with unique global random attractors. In fact, we only need to verify these statements for Hopf bifurcation with additive noise, and by using similar statements, we also obtain the same results for pichfork bifurcation with additive noise.

Consider the two-dimensional stochastic differential equation

$$\begin{aligned}\mathrm{d}x &= (\alpha x - \beta y - (ax - by)(x^2 + y^2))\,\mathrm{d}t + \sigma\,\mathrm{d}W_t^1\,,\\ \mathrm{d}y &= (\alpha y + \beta x - (bx + ay)(x^2 + y^2))\,\mathrm{d}t + \sigma\,\mathrm{d}W_t^2\,,\end{aligned} \tag{5.1}$$

T. S. Doan, *Spectral Theory of Nonautonomous Dynamical Systems and Applications*, https://doi.org/10.1007/978-981-97-5520-2_5

where W_t^1, W_t^2 denote independent one-dimensional Brownian motions.

The stochastic differential equation (5.1) can be rewritten as

$$\mathrm{d}Z_t = f(Z_t)\mathrm{d}t + \sigma \mathrm{d}W_t\,,$$

where $Z_t = (x_t, y_t)^\top$ and $W_t = (\mathrm{d}W_t^1, \mathrm{d}W_t^2)^\top$ and the function $f : \mathbb{R}^2 \to \mathbb{R}^2$ are defined by

$$f(Z) := \begin{pmatrix} \alpha & -\beta \\ \beta & \alpha \end{pmatrix} Z - (x^2 + y^2) \begin{pmatrix} a & -b \\ b & a \end{pmatrix} Z\,.$$

To investigate sample path properties of the solutions of (5.1), it is convenient to work with the canonical sample path space of Brownian motions. Let $\Omega = C_0(\mathbb{R}, \mathbb{R}^2)$ be the space of all continuous functions $\omega : \mathbb{R} \to \mathbb{R}^2$ satisfying that $\omega(0) = 0$. We endow Ω with the compact open topology and denote by $\mathcal{F} = \mathcal{B}(\Omega)$ the Borel σ-algebra on Ω.

It is well known that there exists the so-called *Wiener probability measure* $\mathbb{P}$ on $(\Omega, \mathcal{F})$ which ensures that the two processes $(W_t^1)_{t\in\mathbb{R}}$ and $(W_t^2)_{t\in\mathbb{R}}$, defined by $(W_t^1(\omega), W_t^2(\omega))^\top := \omega(t)$ for $\omega \in \Omega$, are independent two-dimensional Brownian motions. We define the sub σ-algebra $\mathcal{F}_{s,t}$ as the σ-algebra generated by $\omega(u) - \omega(v)$ for $s \le v \le u \le t$. For each $t \in \mathbb{R}$, we define the shift map $\theta_t : \Omega \to \Omega$ by

$$(\theta_t\omega)(s) = \omega(s + t) - \omega(t) \quad \text{for all } s \in \mathbb{R}\,.$$

It is well known that $(\theta_t)_{t\in\mathbb{R}}$ is an ergodic flow preserving the probability measure $\mathbb{P}$; see e.g., [6]. Thus, $(\Omega, \mathcal{F}, \mathbb{P}, (\theta_t)_{t\in\mathbb{R}})$ is an ergodic metric dynamical system; see Definition 3.1.

Given $\omega \in \Omega$, an initial value $Z \in \mathbb{R}^2$ and $T > 0$, we say that a continuous function $\varphi(\cdot, \omega, Z) : [0, T] \to \mathbb{R}^2$ solves the stochastic differential equation (5.1) if it satisfies the integral equation

$$\varphi(t, \omega, Z) = Z + \int_0^t f(\varphi(s, \omega, Z))\,\mathrm{d}s + \sigma\omega(t) \quad \text{for all } t \in [0, T]\,.$$

The first result concerns global existence of solutions of (5.1) for almost every sample path, implying that the solutions do not blow up in forward time. We show that the solutions of (5.1) generate a random dynamical system (θ, φ) (see Definition 3.3 and [6, Definition 1.1.1] for a general definition). This means that the $(\mathcal{B}(\mathbb{R}_0^+) \otimes \mathcal{F} \otimes \mathcal{B}(\mathbb{R}^2), \mathcal{B}(\mathbb{R}^2))$-measurable mapping $\varphi : \mathbb{R}_0^+ \times \Omega \times \mathbb{R}^2 \to \mathbb{R}^2$, $(t, \omega, x) \mapsto \varphi(t, \omega, x)$ is a cocycle over θ, i.e., for all $\omega \in \Omega$, $x \in \mathbb{R}^2$ and $t, s \ge 0$

$$\varphi(0, \omega, \cdot) \equiv \mathrm{Id} \quad \text{and} \quad \varphi(t + s, \omega, x) = \varphi(t, \theta_s\omega, \varphi(s, \omega, x)).$$

In addition to the generation of a random dynamical system, the main theorem of this section also addresses also the existence of a random attractor. For this purpose, we first recall some fundamental facts of random attractor.

A random variable $R : \Omega \to \mathbb{R}$ is called *tempered* if

$$\lim_{t\to\pm\infty} \frac{1}{|t|} \ln^+ R(\theta_t\omega) = 0 \quad \text{for almost all } \omega \in \Omega\,,$$

see also [6, p. 164]. A set $D \in \mathcal{F} \otimes \mathcal{B}(\mathbb{R}^d)$ is called *tempered* if there exists a tempered random variable R such that

$$D(\omega) \subset B_{R(\omega)}(0) \quad \text{for almost all } \omega \in \Omega\,,$$

where $D(\omega) := \{x \in \mathbb{R}^d : (\omega, x) \in D\}$. D is called compact if $D(\omega) \subset \mathbb{R}^d$ is compact for almost all $\omega \in \Omega$. Denote by $\mathcal{D}$ the set of all compact tempered sets $D \in \mathcal{F} \otimes \mathcal{B}(\mathbb{R}^d)$. We now define the notion of a random attractor with respect to $\mathcal{D}$, see also [83, Definition 14.3].

Definition 5.1 (Random Attractor) A set $A \in \mathcal{D}$ is called a *random attractor* (with respect to $\mathcal{D}$) if the following two properties are satisfied.

(i) A is φ-invariant, i.e.,

$$\varphi(t,\omega)A(\omega) = A(\theta_t\omega) \quad \text{for all } t \geq 0 \text{ and almost all } \omega \in \Omega\,.$$

(ii) For all $D \in \mathcal{D}$, we have

$$\lim_{t\to\infty} \operatorname{dist}\big(\varphi(t,\theta_{-t}\omega)D(\theta_{-t}\omega), A(\omega)\big) = 0 \quad \text{for almost all } \omega \in \Omega\,,$$

where $\operatorname{dist}(E,F) := \sup_{x\in E} \inf_{y\in F} \|x - y\|$.

Note that the random attractor is required to be measurable with respect to $\mathcal{F} \otimes \mathcal{B}(\mathbb{R}^d)$, in contrast to a weaker statement normally used in the literature (see also [51, Remark 4]).

The existence of random attractors is proved via so-called absorbing sets. A set $B \in \mathcal{D}$ is called an *absorbing set* if for almost all $\omega \in \Omega$ and any $D \in \mathcal{D}$, there exists a $T > 0$ such that

$$\varphi(t,\theta_{-t}\omega)D(\theta_{-t}\omega) \subset B(\omega) \quad \text{for all } t \geq T\,.$$

A proof of the following theorem can be found in [64, Theorem 3.5].

Theorem 5.1 (Existence of Random Attractors) *Suppose that (θ, φ) is a continuous random dynamical system with an absorbing set B. Then there exists a unique random attractor A, given by*

$$A(\omega) := \bigcap_{\tau \geq 0} \overline{\bigcup_{t \geq \tau} \varphi(t, \theta_{-t}\omega) B(\theta_{-t}\omega)} \quad \textit{for almost all } \omega \in \Omega.$$

Furthermore, $\omega \mapsto A(\omega)$ is measurable with respect to $\mathcal{F}_{-\infty}^0$, i.e., the past of the system.

Now, we are in a position to state the main result of this subsection about generation of a random dynamical system with a random attractor for (5.1).

Theorem 5.2 (Generation of a Random Dynamical System with a Random Attractor) *For the stochastic differential equation* (5.1), *there exists a θ-invariant $\mathcal{F}$-measurable set $\widehat{\Omega} \subset \Omega$ of full probability such that the following statements hold.*

(i) For all $\omega \in \widehat{\Omega}$ and $Z \in \mathbb{R}^2$, the stochastic differential equation (5.1) *admits a unique solution $\varphi(\cdot, \omega, Z)$ such that φ forms a cocycle for a random dynamical system on $(\widehat{\Omega}, \mathcal{F}, \mathbb{P}, (\theta_t)_{t\in\mathbb{R}})$.*
(ii) There exists a random attractor $A \in \mathcal{F}\otimes\mathcal{B}(\mathbb{R}^2)$ of the random dynamical system (θ, φ) such that $\omega \mapsto A(\omega)$ is measurable with respect to $\mathcal{F}_{-\infty}^0$, i.e., the past of the system.

Remark 5.1 Since the difference of the spaces Ω and $\widehat{\Omega}$ is a set of measure zero, we identify both in the following.

We prove Theorem 5.2 by following methods developed in [76, 77]. We conjugate the stochastic differential equation (5.1) to a random differential equation via a suitable transformation using an Ornstein–Uhlenbeck process, so that we need to prove the existence of the random dynamical system and its random attractor for the corresponding random differential equation. An advantage in working with random differential equations (in comparison to stochastic differential equations) is that we can work with sample path estimates of solutions.

For $c > 0$, consider the stochastic differential equation

$$\mathrm{d}Z = -cZ\mathrm{d}t + \mathrm{d}W_t\,, \tag{5.2}$$

where $Z \in \mathbb{R}^2$. Define the random variable $Z^* := \int_{-\infty}^0 e^{cs}\mathrm{d}W_s$. Then $t \mapsto Z^*(\theta_t\omega)$ solves (5.2), i.e.,

$$Z^*(\theta_t\omega) = Z^*(\omega) - c\int_0^t Z^*(\theta_s\omega)\,\mathrm{d}s + \omega(t)\,. \tag{5.3}$$

By replacing Ω with a measurable subset $\widehat{\Omega} \subset \Omega$ of full probability that is invariant under θ, there exist two random variables K and L such that

$$|Z^*(\theta_t\omega)|^2 \leq K(\omega) + L(\omega)\ln(1 + |t|) \quad \text{for all } t \in \mathbb{R} \text{ and } \omega \in \widehat{\Omega}\,,$$

see [81]. We define the map $T : \widehat{\Omega} \times \mathbb{R}^2 \to \mathbb{R}^2$ by $T(\omega, Z) := Z + \sigma Z^*(\omega)$. Under the change of variable $Z \mapsto T(\omega, Z)$, the SDE (5.1) is transformed into the random differential equation

$$\dot{Z} = g(\theta_t\omega, Z)\,, \tag{5.4}$$

where $g(\omega, Z) := f(T(\omega, Z)) + c\sigma Z^*(\omega)$. We show later in Lemma 5.2 and the proof of Theorem 5.2 that the solution $\Psi(t, \omega, Z)$ of this random differential equation,

$$\Psi(t, \omega, Z) = Z + \int_0^t g(\theta_s\omega, \Psi(s, \omega, Z))\,\mathrm{d}s\,,$$

exists for all $t \geq 0$ and forms a random dynamical system. The following lemma holds using this fact.

Lemma 5.1 *The following statements hold.*

(i) The random dynamical system $\varphi : \mathbb{R}_0^+ \times \widehat{\Omega} \times \mathbb{R}^2 \to \mathbb{R}^2$*, defined by*

$$\varphi(t, \omega, Z) := T(\theta_t\omega, \Psi(t, \omega, T(\omega)^{-1}Z))\,, \tag{5.5}$$

is generated by the stochastic differential equation (5.1).

(ii) If the random dynamical system Ψ *has a random attractor, then also the random dynamical system* φ *has a random attractor.*

Proof

(i) From (5.5) and the definition of T, we have

$$\varphi(t, \omega, Z) = \Psi(t, \omega, Z - \sigma Z^*(\omega)) + \sigma Z^*(\theta_t\omega)\,,$$

which together with the fact that Ψ is a solution of (5.4) implies that

$$\begin{aligned}\varphi(t, \omega, Z) &= Z - \sigma Z^*(\omega) + \int_0^t g(\theta_s\omega, \Psi(s, \omega, Z - \sigma Z^*(\omega)))\,\mathrm{d}s \\ &\quad + \sigma Z^*(\theta_t\omega) \\ &= Z + \int_0^t g(\theta_s\omega, T(\theta_s\omega)^{-1}(\varphi(s, \omega, Z)))\,\mathrm{d}s \\ &\quad + \sigma(Z^*(\theta_t\omega) - Z^*(\omega))\,.\end{aligned}$$

Thus, using (5.3), we obtain that

$$\varphi(t, \omega, Z) = Z + \int_0^t g(\theta_s\omega, T(\theta_s\omega)^{-1}(\varphi(s, \omega, Z)))$$

$$
\begin{aligned}
&- \sigma c Z^*(\theta_s\omega)\,\mathrm{d}s + \sigma\omega(t) \\
&= Z + \int_0^t f(\theta_s\omega, \varphi(s,\omega,Z))\,\mathrm{d}s + \sigma\omega(t)\,,
\end{aligned}
$$

which completes the proof of this part.

(ii) This follows from the definition of a random attractor and the fact that the shifted term in the transformation $T(\omega, Z)$, namely, $Z^*(\theta_t\omega)$, is tempered.

We show that the Euclidean norm of the solutions of (5.4) is bounded by the growth of the corresponding solutions of the scalar equation

$$
\dot{\zeta} = \gamma_t(\omega) - \sqrt{a}\zeta\,, \tag{5.6}
$$

where the stochastic process $(\gamma_t)_{t\in\mathbb{R}}$ is chosen appropriately. Note that for each initial value $\zeta_0 \in \mathbb{R}$, the explicit solution of (5.6) is given by

$$
\zeta(t,\omega,\zeta_0) = e^{-\sqrt{a}t}\zeta_0 + \int_0^t e^{-\sqrt{a}(t-s)}\gamma_s(\omega)\,\mathrm{d}s\,. \tag{5.7}
$$

Lemma 5.2 *There exists a tempered stochastic processes $(\gamma_t)_{t\in\mathbb{R}}$, i.e.,*

$$
\lim_{t\to\pm\infty} \frac{|\gamma_t(\omega)|}{e^{\epsilon|t|}} = 0 \quad \text{for all } \epsilon > 0 \text{ and } \omega \in \widehat{\Omega}\,,
$$

such that for $Z \in \mathbb{R}^2$, we have

$$
\|\Psi(t,\omega,Z)\|^2 \le 2\zeta(t,\omega,\|Z\|^2)\,, \tag{5.8}
$$

which implies that the solution $\Psi(t,\omega,Z)$ exists for all $t \ge 0$.

Proof By replacing Z with $(x,y)^\top$ and Z^* with $(x^*,y^*)^\top$, we rewrite (5.4) as

$$
\begin{aligned}
\begin{pmatrix} \dot{x}_t \\ \dot{y}_t \end{pmatrix} &= \begin{pmatrix} \alpha & -\beta \\ \beta & \alpha \end{pmatrix} \begin{pmatrix} x_t + \sigma x^*(\theta_t\omega) \\ y_t + \sigma y^*(\theta_t\omega) \end{pmatrix} + c\sigma \begin{pmatrix} x^*(\theta_t\omega) \\ y^*(\theta_t\omega) \end{pmatrix} \\
&\quad - \left\| \begin{pmatrix} x_t + \sigma x^*(\theta_t\omega) \\ y_t + \sigma y^*(\theta_t\omega) \end{pmatrix} \right\|^2 \begin{pmatrix} a & -b \\ b & a \end{pmatrix} \begin{pmatrix} x_t + \sigma x^*(\theta_t\omega) \\ y_t + \sigma y^*(\theta_t\omega) \end{pmatrix}.
\end{aligned}
$$

Let $r_t := \frac{1}{2}(x_t^2 + y_t^2)$. Then a direct computation yields that

$$
\begin{aligned}
\dot{r}_t &= x_t\dot{x}_t + y_t\dot{y}_t \\
&= 2\alpha r_t + \sigma x^*(\theta_t\omega)((\alpha + c)x_t + \beta y_t) - \sigma y^*(\theta_t\omega)(\beta x_t - (\alpha + c)y_t)
\end{aligned}
$$

$$
-\left\|\begin{pmatrix} x_t+\sigma x^*(\theta_t\omega) \\ y_t+\sigma y^*(\theta_t\omega) \end{pmatrix}\right\|^2 \Big(2ar_t+\sigma x^*(\theta_t\omega)(ax_t+by_t)
$$
$$
-\sigma y^*(\theta_t\omega)(bx_t-ay_t)\Big)\,.
$$

Note that $\max\{(\alpha+c)x_t+\beta y_t, \beta x_t-(\alpha+c)y_t\} \le \sqrt{((\alpha+c)^2+\beta^2)2r_t}$. Thus,

$$
\begin{aligned}
&|x^*(\theta_t\omega)((\alpha+c)x_t+\beta y_t)-y^*(\theta_t\omega)(\beta x_t-(\alpha+c)y_t)| \\
&\le \sqrt{((\alpha+c)^2+\beta^2)2r_t}\,(|x^*(\theta_t\omega)|+|y^*(\theta_t\omega)|) \\
&\le 2\sqrt{((\alpha+c)^2+\beta^2)r_t}\,\|Z^*(\theta_t\omega)\|\,.
\end{aligned}
\tag{5.9}
$$

On the other hand, we have

$$
\left\|\begin{pmatrix} x_t+\sigma x^*(\theta_t\omega) \\ y_t+\sigma y^*(\theta_t\omega) \end{pmatrix}\right\|^2 = 2r_t+\sigma^2\|Z^*(\theta_t\omega)\|^2+2\sigma x^*(\theta_t\omega)x_t+2\sigma y^*(\theta_t\omega)y_t\,,
$$

which together with the fact that $|x^*(\theta_t\omega)x_t+y^*(\theta_t\omega)y_t| \le \|Z^*(\theta_t\omega)\|\sqrt{2r_t}$ implies that

$$
\left|\left\|\begin{pmatrix} x_t+\sigma x^*(\theta_t\omega) \\ y_t+\sigma y^*(\theta_t\omega) \end{pmatrix}\right\|^2 - 2r_t-\sigma^2\|Z^*(\theta_t\omega)\|^2\right| \le 2\sigma\|Z^*(\theta_t\omega)\|\sqrt{2r_t}\,.
$$

Consequently,

$$
ar_t\left\|\begin{pmatrix} x_t+\sigma x^*(\theta_t\omega) \\ y_t+\sigma y^*(\theta_t\omega) \end{pmatrix}\right\|^2 \ge 2ar_t^2-2^{\frac{3}{2}}a\sigma\|Z^*(\theta_t\omega)\|r_t^{\frac{3}{2}}+a\sigma^2\|Z^*(\theta_t\omega)\|^2r_t\,,
\tag{5.10}
$$

and from the fact that

$$
|x^*(\theta_t\omega)(ax_t+by_t)-y^*(\theta_t\omega)(bx_t-ay_t)| \le \sqrt{a^2+b^2}\|Z^*(\theta_t\omega)\|\sqrt{2r_t}\,,
$$

we derive that

$$
\begin{aligned}
&\Big|\sigma x^*(\theta_t\omega)(ax_t+by_t)-\sigma y^*(\theta_t\omega)(bx_t-ay_t)\Big|\left\|\begin{pmatrix} x_t+\sigma x^*(\theta_t\omega) \\ y_t+\sigma y^*(\theta_t\omega) \end{pmatrix}\right\|^2 \\
&\le 2^{\frac{3}{2}}\sigma\sqrt{a^2+b^2}\|Z^*(\theta_t\omega)\|r_t^{\frac{3}{2}}+\sigma^2\sqrt{2(a^2+b^2)}\|Z^*(\theta_t\omega)\|^2r_t \\
&+\sigma^3\sqrt{2(a^2+b^2)}\|Z^*(\theta_t\omega)\|^3\sqrt{r_t}\,.
\end{aligned}
\tag{5.11}
$$

Using (5.9), (5.10), (5.11) and a comparison argument, we obtain for all $t \ge 0$ and $Z \in \mathbb{R}^2 \setminus \{0\}$ that $\frac{1}{2}\|\Psi(t,\omega,Z)\|^2 \le \tilde{\zeta}(t,\omega,\|Z\|^2)$, where $t \mapsto \tilde{\zeta}(t,\omega,\|Z\|^2) = \tilde{\zeta}_t$ is the solution of the following scalar differential equation

$$\dot{\tilde{\zeta}}_t = a_t(\omega)\tilde{\zeta}_t^{\frac{1}{2}} + b_t(\omega)\tilde{\zeta}_t + c_t(\omega)\tilde{\zeta}_t^{\frac{3}{2}} - 4a\tilde{\zeta}_t^2 ,$$

with initial condition $\tilde{\zeta}_0 = \|Z\|^2$. Here, the functions a_t, b_t, c_t are defined by

$$\begin{aligned}
a_t(\omega) &:= 2\sigma\sqrt{(\alpha+c)^2+\beta^2}\|Z^*(\theta_t\omega)\| + \sqrt{2}\sigma^3\sqrt{a^2+b^2}\|Z^*(\theta_t\omega)\|^2 ,\\
b_t(\omega) &:= 2\alpha + 4\sigma^2\sqrt{a^2+b^2}\|Z^*(\theta_t\omega)\|^2 - 2a\sigma^2\|Z^*(\theta_t\omega)\|^2 ,\\
c_t(\omega) &:= \left(2^{3/2}\sqrt{a^2+b^2}\sigma + 2^{5/2}a\sigma\right)\|Z^*(\theta_t\omega)\| .
\end{aligned}$$

From temperedness of $Z^*(\theta_t\omega)$, all stochastic processes $(a_t)_{t\in\mathbb{R}}$, $(b_t)_{t\in\mathbb{R}}$ and $(c_t)_{t\in\mathbb{R}}$ are also tempered. Note that

$$\begin{aligned}
a\tilde{\zeta}_t^2 + \sqrt[3]{\frac{a_t(\omega)^4}{4^4 a}} &\geq |a_t(\omega)|\tilde{\zeta}_t^{\frac{1}{2}} ,\\
a\tilde{\zeta}_t^2 + \frac{b_t(\omega)^2}{2a} &\geq |b_t(\omega)\tilde{\zeta}_t| ,\\
a\tilde{\zeta}_t^2 + \frac{3^3 c_t(\omega)^4}{4^4 a^3} &\geq |c_t(\omega)\tilde{\zeta}_t^{\frac{3}{2}}| .
\end{aligned}$$

Therefore,

$$\begin{aligned}
&a_t(\omega)\tilde{\zeta}_t^{\frac{1}{2}} + b_t(\omega)\tilde{\zeta}_t + c_t(\omega)\tilde{\zeta}_t^{\frac{3}{2}} - 4a\tilde{\zeta}_t^2\\
&\leq \sqrt[3]{\frac{a_t(\omega)^4}{4^4 a}} + \frac{b_t(\omega)^2}{2a} + \frac{3^3 c_t(\omega)^4}{4^4 a^3} - a\tilde{\zeta}_t^2\\
&\leq \gamma_t(\omega) - \sqrt{a}\tilde{\zeta}_t ,
\end{aligned}$$

where

$$\gamma_t(\omega) := \frac{1}{4} + \sqrt[3]{\frac{a_t(\omega)^4}{4^4 a}} + \frac{b_t(\omega)^2}{2a} + \frac{3^3 c_t(\omega)^4}{4^4 a^3}$$

is tempered. Hence, using a comparison argument, the solution ζ of (5.6) satisfies (5.8), which finishes the proof of this lemma.

Proof of Theorem 5.2

(i) According to [6], there exists a local random dynamical system generated by solutions of (5.4). Due to Lemma 5.2, the solution $\Psi(t, \omega, Z)$ exists for all $t \geq 0$. Hence, this proves the fact that we assumed to prove Lemma 5.1. Lemma 5.1 (i) completes the proof of (i).

(ii) Let $D \in \mathcal{F} \otimes \mathcal{B}(\mathbb{R}^d)$ be tempered. Then there exists a tempered random variable $R : \widehat{\Omega} \to \mathbb{R}^+$ such that $D(\omega) \subset B_{R(\omega)}(0)$. By Lemma 5.2, for all $Z \in D(\theta_{-t}\omega)$, we have

$$\begin{aligned}\|\Psi(t, \theta_{-t}\omega, Z)\|^2 &\leq 2\zeta(t, \theta_{-t}\omega, R(\theta_{-t}\omega)) \\ &\leq 2e^{-\sqrt{a}t} R(\theta_{-t}\omega) + 2\int_{-t}^{0} e^{\sqrt{a}s} \gamma_s(\omega)\, \mathrm{d}s\,,\end{aligned}$$

where we use (5.7) to obtain the last inequality. Since $(\gamma_t)_{t \in \mathbb{R}}$ is tempered, $\int_{-\infty}^{0} e^{\sqrt{a}s} \gamma_s(\omega)\, \mathrm{d}s$ exists. On the other hand, since R is tempered, it follows that $\lim_{t \to \infty} e^{-\sqrt{a}t} R(\theta_{-t}\omega) = 0$. Define $r(\omega) := \sqrt{1 + 2\int_{-\infty}^{0} e^{\sqrt{a}s} \gamma_s(\omega)\, \mathrm{d}s}$. Thus, for each $\omega \in \widehat{\Omega}$, there exists $T > 0$ such that

$$\Psi(t, \theta_{-t}\omega, D(\theta_{-t}\omega)) \subset B_{r(\omega)}(0) \quad \text{for all } t \geq T\,.$$

This means that $B_{r(\omega)}(0)$ is an absorbing set. Applying Theorem 5.1 completes the proof.

5.2 Random Pitchfork Bifurcation

This section is devoted to analyzing the following model of *pitchfork bifurcation with additive noise*

$$\mathrm{d}x = \big(\alpha x - x^3\big)\mathrm{d}t + \sigma \mathrm{d}W_t\,, \tag{5.12}$$

depending on real parameters α and σ and driven by a two-sided Wiener process $(W_t)_{t \in \mathbb{R}}$. We first look at the deterministic case $\sigma = 0$. Then for $\alpha < 0$, the ordinary differential equation (5.12) has one equilibrium ($x = 0$) which is globally attractive. For positive α, the trivial equilibrium becomes repulsive, and there are two additional equilibria, given by $\pm\sqrt{\alpha}$, which are attractive. This also means that the global attractor K_α of the deterministic equation undergoes a bifurcation from a trivial to a nontrivial object. It is given by

$$K_\alpha := \begin{cases} \{0\} & : \alpha \leq 0\,, \\ \big[-\sqrt{\alpha}, \sqrt{\alpha}\big] & : \alpha > 0\,. \end{cases}$$

It was shown in [50] that such an attractor bifurcation does not persist for random attractors of the randomly perturbed system where $\sigma > 0$, and we will explain the details now.

Firstly, by analyzing in Sect. 5.1, the stochastic differential equation (5.12) generates a random dynamical system $(\theta : \mathbb{R} \times \Omega \to \Omega, \varphi : \mathbb{R}_{\geq 0} \times \Omega \times \mathbb{R} \to \mathbb{R})$ in the sense that $\Omega = C_0(\mathbb{R}, \mathbb{R})$ be the space of all continuous functions $\omega : \mathbb{R} \to \mathbb{R}$ satisfying that $\omega(0) = 0$ and $\varphi(t, \omega, x)$ denote the solution at time t of (5.12) which satisfies that $x(0) = x$.

The random dynamical system φ also generates a Markov semigroup with transition probabilities $T(x, B)$ for $x \in \mathbb{R}$ and $B \in \mathcal{B}(\mathbb{R})$. A probability measure ρ on $\mathcal{B}(X)$ is called a *stationary measure* for the Markov semigroup if

$$\rho(B) = \int_{\mathbb{R}} T(x, B)\, \mathrm{d}\rho(x) \quad \text{for all } B \in \mathcal{B}(\mathbb{R}).$$

It can be shown [6, p. 474] that for any $\alpha, \sigma \in \mathbb{R}$, the Markov semigroup associated with (5.12) admits a unique stationary measure $\rho_{\alpha,\sigma}$ with density

$$p_{\alpha,\sigma}(x) = N_{\alpha,\sigma} \exp\left(\tfrac{1}{\sigma^2}(\alpha x^2 - \tfrac{1}{2}x^4)\right), \tag{5.13}$$

where $N_{\alpha,\sigma}$ is a normalization constant. This stationary measure corresponds to an invariant measure μ of the random dynamical system (θ, φ) generated by (5.12). μ has the disintegration given by

$$\mu_\omega = \lim_{t\to\infty} \varphi(t, \theta_{-t}\omega)\rho \quad \text{for almost all } \omega \in \Omega.$$

It was shown in [50] that μ_ω is a Dirac measure concentrated on $a_\alpha(\omega)$, and linearizing along this invariant measure μ yields a negative Lyapunov exponent, given by

$$\lambda_\alpha = -\frac{2}{\sigma^2} \int_{\mathbb{R}} (\alpha x - x^3)^2 p_{\alpha,\sigma}(x)\, \mathrm{d}x.$$

Moreover, the family $\{a_\alpha(\omega)\}_{\omega\in\Omega}$ is the global random attractor which implies that the attractor bifurcation associated with a deterministic pitchfork bifurcation (i.e., K_α bifurcates from a nontrivial object to a singleton) is destroyed by noise. However, in what follows, we show in Sect. 5.2.1 that there is a qualitative change in the random dynamics at the bifurcation point $\alpha = 0$ in the sense that after the bifurcation, the attracting random fixed points $\{a_\alpha(\omega)\}_{\omega\in\Omega}$ have qualitatively different properties for $\alpha < 0$ and $\alpha \geq 0$ with respect to uniform attraction, which is lost at the bifurcation point. We also associate this bifurcation in Sect. 5.2.2 with the dichotomy spectrum of the linearization at the bifurcation point.

5.2.1 Qualitative Changes in Uniform Attractivity

In order to establish qualitative changes in the attractivity of the unique random attracting fixed point $\{a_\alpha(\omega)\}_{\omega\in\Omega}$, a detailed understanding about the location of this attractor is needed. For this purpose, we use similar techniques as developed in [119, 120].

Proposition 5.1 *Consider* (5.12) *for* $\alpha \in \mathbb{R}$*, and let* $\{a_\alpha(\omega)\}_{\omega\in\Omega}$ *be its unique random fixed point. Then for any* $\varepsilon > 0$ *and* $T \geq 0$*, there exists a measurable set* $\mathcal{A} \in \mathcal{F}_{-\infty}^T$ *of positive measure such that*

$$a_\alpha(\theta_s\omega) \in (-\varepsilon, \varepsilon) \quad \textit{for all } s \in [0, T] \textit{ and } \omega \in \mathcal{A}\,.$$

Proof The unique stationary measure $\rho_{\alpha,\sigma}$ for the Markov semigroup associated with (5.12) with $|\sigma| > 0$ is equivalent to the Lebesgue measure with the density function given by (5.13). The invariant measure $\delta_{a(\omega)}$ and stationary measure ρ are in correspondence by the following relations: The invariant measure is obtained as the limit of the pullback images of the stationary measure, i.e.,

$$\delta_{a(\omega)} = \lim_{t\to\infty} \varphi(t, \theta_{-t}\omega)\rho \quad \text{for almost all } \omega \in \Omega\,,$$

and the stationary measure is obtained as the expectation of the invariant measure, i.e.,

$$\rho(\cdot) = \int_\Omega \delta_{a(\omega)}(\cdot)\mathrm{d}\mathbb{P}(\omega) \tag{5.14}$$

(see [50]). Now define

$$\eta := \frac{\varepsilon e^{-|\alpha|T}}{2(1+|\sigma|)}.$$

Since the support of ρ is the entire real line, it follows from (5.14) that the set

$$A_1 := \{\omega \in \Omega\colon a_\alpha(\omega) \in (-\eta, \eta)\}$$

has positive probability for any $\alpha \in \mathbb{R}$. The global pullback attractor $\{a_\alpha(\omega)\}$ is measurable with respect to the past of the noise $\mathcal{F}_{-\infty}^0$ (again see [50]), and hence $A_1 \in \mathcal{F}_{-\infty}^0$. Define

$$A_2 := \left\{\omega \in \Omega\colon \sup\nolimits_{t\in[0,T]} |\omega(t)| \leq \tfrac{\eta}{2}\right\} \in \mathcal{F}_0^T$$

which, by Ikeda and Watanabe [73, Section 6.8], has positive probability. Since the sets A_1 and A_2 are independent, the set $\mathcal{A} := A_1 \cap A_2 \in \mathcal{F}_{-\infty}^T$ also has positive probability. Choose and fix an arbitrary $\omega \in \mathcal{A}$. By the definition of A_1, we have that $|a_\alpha(\omega)| < \eta$. Since $a_\alpha(\omega)$ is a random fixed point of φ, it follows, using the integral from of (5.12), that

$$a_\alpha(\theta_t\omega) = a_\alpha(\theta_s\omega) + \int_s^t \left(\alpha a_\alpha(\theta_r\omega) - a_\alpha(\theta_r\omega)^3\right) \mathrm{d}r + \sigma(\omega(t) - \omega(s)) . \qquad (5.15)$$

Choose and fix an arbitrary $t \in [0, T]$. Define $\mathcal{I} := \{s \in [0, t] : a_\alpha(\theta_s\omega) = 0\}$; by continuity, the set $\mathcal{I}$ is closed (but possibly empty). We consider the following three cases:
Case 1. If $t \in \mathcal{I}$, then $|a_\alpha(\theta_t\omega)| = 0$.
Case 2. If $\mathcal{I}$ is not empty and $t \notin \mathcal{I}$, then $s := \sup \mathcal{I} < t$ and $a_\alpha(\theta_s\omega) = 0$. By the definition of $\mathcal{I}$ and continuity, we have either $a_\alpha(\theta_r\omega) > 0$ for all $r \in (s, t]$ or $a_\alpha(\theta_r\omega) < 0$ for all $r \in (s, t]$. Using this observation and (5.15), we obtain that

$$|a_\alpha(\theta_t\omega)| \le |\sigma|\eta + \int_s^t |\alpha||a_\alpha(\theta_r\omega)| \,\mathrm{d}r.$$

Case 3. If $\mathcal{I}$ is empty, then either $a_\alpha(\theta_s\omega) > 0$ for all $s \in [0, t]$ or $a_\alpha(\theta_s\omega) < 0$ for all $s \in [0, t]$. Using (5.15) and noting that $|a_\alpha(\omega)| < \eta$, we arrive at the following inequality:

$$|a_\alpha(\theta_t\omega)| \le (1 + |\sigma|)\eta + \int_0^t |\alpha||a_\alpha(\theta_s\omega)| \,\mathrm{d}s.$$

In view of the three cases above, we have that

$$|a_\alpha(\theta_t\omega)| \le (1 + |\sigma|)\eta + \int_0^t |\alpha||a_\alpha(\theta_s\omega)| \,\mathrm{d}s \quad \text{for all } t \in [0, T] .$$

Then, using Gronwall's inequality, we obtain that

$$|a_\alpha(\theta_t\omega)| \le (1 + |\sigma|)\eta e^{|\alpha|t} < \varepsilon \quad \text{for all } t \in [0, T] .$$

Thus we have that for all $\omega \in \mathcal{A}$, $a_\alpha(\theta_t\omega) \in (-\varepsilon, \varepsilon)$ for all $t \in [0, T]$, which completes the proof.

We now give a detailed description of the random bifurcation scenario for the stochastic differential equation (5.12) by means of both *asymptotic* and *finite-time* dynamical behavior. The asymptotic description implies that there is a qualitative change in the uniformity of attraction of the unique random attractor $\{a_\alpha(\omega)\}_{\omega\in\Omega}$. On the other hand, the finite-time description shows that after the bifurcation, even if the time interval is very large, the (asymptotic) Lyapunov exponent cannot be observed with nonvanishing probability (by a finite-time Lyapunov exponent); however, before the bifurcation, the (asymptotic) Lyapunov exponent can be approximated by the finite-time Lyapunov exponent. Finite-time Lyapunov exponents for random dynamical systems have not been considered in the literature so far but play an important role in the description of Lagrangian coherent structures in fluid dynamics [69].

Let $\{a_\alpha(\omega)\}_{\omega\in\Omega}$ denote the unique random attracting fixed point of a stochastic differential equation (5.12). Then $\{a_\alpha(\omega)\}_{\omega\in\Omega}$ is called *locally uniformly attractive* if there exists $\delta > 0$ such that

$$\lim_{t\to\infty} \sup_{x\in(-\delta,\delta)} \operatorname*{ess\,sup}_{\omega\in\Omega} |\varphi(t,\omega)(a_\alpha(\omega)+x) - a_\alpha(\theta_t\omega)| = 0.$$

Theorem 5.3 (Random Pitchfork Bifurcation, Asymptotic Description) *Consider the stochastic differential equation* (5.12) *with the unique random attracting fixed point* $\{a_\alpha(\omega)\}_{\omega\in\Omega}$. *Then the following statements hold:*

(i) For $\alpha < 0$, *the random attractor* $\{a_\alpha(\omega)\}_{\omega\in\Omega}$ *is locally uniformly attractive; in fact, it is even globally uniformly exponential attractive, i.e.,*

$$|\varphi(t,\omega,x) - \varphi(t,\omega,a_\alpha(\omega))| \le e^{\alpha t}|x - a_\alpha(\omega)| \quad \textit{for all } x \in \mathbb{R}. \tag{5.16}$$

(ii) For $\alpha > 0$, *the random attractor* $\{a_\alpha(\omega)\}_{\omega\in\Omega}$ *is not locally uniformly attractive.*

Proof

(i) Let $x \in \mathbb{R}$ be arbitrary such that $x \neq a_\alpha(\omega)$. Using the monotonicity of solutions, we may assume that $\varphi(t,\omega,x) > \varphi(t,\omega,a_\alpha(\omega))$ for all $t \ge 0$. The integral form of (5.12),

$$\varphi(t,\omega)x = x + \int_0^t \left(\alpha\varphi(s,\omega)x - (\varphi(s,\omega)x)^3\right) \mathrm{d}s + \sigma\omega(t)$$

yields that

$$\varphi(t,\omega)x - \varphi(t,\omega)a_\alpha(\omega) \le x - a_\alpha(\omega) + \alpha\int_0^t \left(\varphi(s,\omega)x - \varphi(s,\omega)a_\alpha(\omega)\right) \mathrm{d}s\,.$$

Using Gronwall's inequality implies (5.16), which finished this part of the proof.

(ii) Suppose to the contrary that there exists $\delta > 0$ such that

$$\lim_{t\to\infty} \sup_{x\in(-\delta,\delta)} \operatorname*{ess\,sup}_{\omega\in\Omega} |\varphi(t,\omega,a_\alpha(\omega)+x) - a_\alpha(\theta_t\omega)| = 0\,,$$

which implies that there exists $N \in \mathbb{N}$ such that

$$\sup_{x\in(-\delta,\delta)} \operatorname*{ess\,sup}_{\omega\in\Omega} |\varphi(t,\omega,a_\alpha(\omega)+x) - a_\alpha(\theta_t\omega)| < \frac{\sqrt{\alpha}}{4} \quad \text{for all } t \ge N\,. \tag{5.17}$$

According to Proposition 5.1, there exists $\mathcal{A} \in \mathcal{F}_{-\infty}^{0}$ of positive probability such that $a_\alpha(\omega) \in (-\frac{\delta}{4}, \frac{\delta}{4})$. Note that $-\sqrt{\alpha}$ and $\sqrt{\alpha}$ are two attractive fixed points for the deterministic differential equation

$$\dot{x} = \alpha x - x^3 .$$

Let $\varphi(\cdot, x_0)$ denote the solution of the above deterministic equation which satisfies that $x(0) = x_0$. Then there exists $T > N$ such that

$$\varphi(T, \delta/4) > \frac{\sqrt{\alpha}}{2} \quad \text{and} \quad \varphi(T, -\delta/4) < -\frac{\sqrt{\alpha}}{2} . \tag{5.18}$$

For any $\varepsilon > 0$, we define

$$\mathcal{A}_\varepsilon^+ := \left\{\omega \in \Omega : \sup_{t\in[0,T]} |\omega(t)| < \varepsilon\right\} .$$

Clearly, $\mathcal{A}_\varepsilon^+ \in \mathcal{F}_0^T$ has positive probability, and thus, $\mathbb{P}(\mathcal{A}\cap\mathcal{A}_\varepsilon^+) = \mathbb{P}(\mathcal{A})\mathbb{P}(\mathcal{A}_\varepsilon^+)$ is positive. Due to compactness of $[0, T]$, there exists $\varepsilon > 0$ such that for all $\omega \in \mathcal{A}_\varepsilon^+$, we have

$$|\varphi(T, \omega, \delta/4) - \varphi(T, \delta/4)| \le \frac{\sqrt{\alpha}}{4} \text{ and } |\varphi(T, \omega, -\delta/4) - \varphi(T, -\delta/4)| < \frac{\sqrt{\alpha}}{4} ,$$

which implies together with (5.18) that

$$\varphi(T, \omega, \delta/4) > \frac{\sqrt{\alpha}}{4} \quad \text{and} \quad \varphi(T, \omega, -\delta/4) < -\frac{\sqrt{\alpha}}{4} .$$

Due to the fact that $|a_\alpha(\omega)| \le \frac{\delta}{2}$ for all $\omega \in \mathcal{A} \cap \mathcal{A}_\varepsilon^+$, we get that for all $\omega \in \mathcal{A} \cap \mathcal{A}_\varepsilon^+$

$$\begin{aligned} &\sup_{x\in(-\delta,\delta)} |\varphi(T, \omega, a_\alpha(\omega) + x) - a_\alpha(\theta_T\omega)| \\ &\ge \max\left\{\varphi(T, \omega, \delta/4) - a_\alpha(\theta_T\omega)|, |\varphi(T, \omega, -\delta/4) - a_\alpha(\theta_T\omega)|\right\} . \end{aligned}$$

Consequently,

$$\sup_{x\in(-\delta,\delta)} \operatorname*{ess\,sup}_{\omega\in\Omega} |\varphi(t, \omega, a_\alpha(\omega) + x) - a_\alpha(\theta_t\omega)| > \frac{\sqrt{\alpha}}{4} ,$$

which contradicts to (5.17) and the proof is complete.

For the description of the bifurcation via finite-time properties, consider a compact time interval $I = [0, T]$, and define the corresponding *finite-time Lyapunov exponent* associated with the invariant measure $a_\alpha(\omega)$ by

$$\lambda_\alpha^{T,\omega} := \frac{1}{T} \ln \left| \frac{\partial \varphi_\alpha}{\partial x}(T, \omega, a_\alpha(\omega)) \right| .$$

Clearly, the (classical) Lyapunov exponent λ_α^∞ associated with the random fixed point $a_\alpha(\omega)$ is given by

$$\lambda_\alpha^\infty = \lim_{T\to\infty} \lambda_\alpha^{T,\omega} .$$

In contrast to classical Lyapunov exponent, the finite-time Lyapunov exponent is, in general, a nonconstant random variable.

Theorem 5.4 (Random Pitchfork Bifurcation, Finite-Time Description) *Consider the stochastic differential equation* (5.12) *with the unique random attracting fixed point* $\{a_\alpha(\omega)\}_{\omega\in\Omega}$. *For any finite time interval* $[0, T]$, *let* $\lambda_\alpha^{T,\omega}$ *denote the finite-time Lyapunov exponent associated with* $\{a_\alpha(\omega)\}_{\omega\in\Omega}$. *Then the following statements hold:*

(i) For $\alpha < 0$, *the random attractor* $\{a_\alpha(\omega)\}_{\omega\in\Omega}$ *is finite-time attractive, i.e.,*

$$\lambda_\alpha^{T,\omega} \le \alpha < 0 \quad \textit{for all } \omega \in \Omega .$$

(ii) For $\alpha > 0$, *the random attractor* $\{a_\alpha(\omega)\}_{\omega\in\Omega}$ *is not finite-time attractive, i.e.,*

$$\mathbb{P}\big\{\omega \in \Omega : \lambda_\alpha^{T,\omega} > 0\big\} > 0.$$

Proof

(i) follows directly from Theorem 5.3 (i).
(ii) Choose $\varepsilon := \frac{\sqrt{\alpha}}{2} > 0$. According to Proposition 5.1, there exists a measurable set $\mathcal{A} \in \mathcal{F}_{-\infty}^T$ of positive probability such that

$$a_\alpha(\theta_s\omega) \in (-\varepsilon, \varepsilon) \quad \text{for all } s \in [0, T] .$$

Let $\omega \in \mathcal{A}$ be arbitrary, and we will estimate $\lambda_\alpha^{T,\omega}$. Note that the linearized equation along the random fixed point $a_\alpha(\omega)$ is given by

$$\dot{\xi}(t) = (\alpha - 3a_\alpha(\theta_t\omega)^2)\xi(t) .$$

We thus get

$$\lambda_\alpha^{T,\omega} = \alpha - \frac{1}{T}\int_0^T 3a_\alpha(\theta_t\omega)^2 \mathrm{d}t \ge \frac{\alpha}{4} ,$$

which completes the proof.

Remark 5.2 For a system without any special structure, finite-time Lyapunov exponents are exact numerically computable quantities which measure expansion rate or contraction rate along a random fixed point. Therefore, the change of the sign of finite-time Lyapunov exponents is definitely a bifurcated point that we observe from a practical point of view.

For a system with some specified structures such as positivity, we refer to [105] for a powerful method to compute the maximal Lyapunov exponent with an explicit bound.

5.2.2 *The Dichotomy Spectrum at the Bifurcation Point*

We will compute the dichotomy spectrum of the linearization around the unique random attracting fixed point $\{a_\alpha(\omega)\}$ of the system (5.12) (see Sect. 3.2.2 for the notion of dichotomy spectrum of linear random dynamical systems and the corresponding spectral theory). As a direct consequence, we observe that hyperbolicity is lost at the bifurcated point $\alpha = 0$.

Theorem 5.5 *Let* $\Phi_\alpha(t,\omega) := \frac{\partial\varphi_\alpha}{\partial x}(t,\omega,a_\alpha(\omega))$ *denote the linearized random dynamical system along the random fixed point* $a_\alpha(\omega)$. *Then the dichotomy spectrum* Σ_α *of* Φ_α *is given by*

$$\Sigma_\alpha = [-\infty, \alpha] \quad \textit{for all } \alpha \in \mathbb{R}\,.$$

Proof From the linearized equation along $a_\alpha(\omega)$

$$\dot{\xi}(t) = (\alpha - 3a_\alpha(\theta_t\omega)^2)\xi(t)\,,$$

we derive that

$$\Phi_\alpha(t,\omega) = \exp\left(\int_0^t \left(\alpha - 3a_\alpha(\theta_s\omega)^2\right)\mathrm{d}s\right)\,. \tag{5.19}$$

Consequently,

$$|\Phi_\alpha(t,\omega)| \le e^{\alpha|t|} \quad \text{for all } t \in \mathbb{R}\,,$$

which implies that $\Sigma_\alpha \subset (-\infty,\alpha]$. Thus, it is sufficient to show that $(-\infty,\alpha] \subset \Sigma_\alpha$. For this purpose, let $\gamma \in (-\infty,\alpha]$ be arbitrary. Suppose the opposite that Φ_α admits an exponential dichotomy with growth rate γ with an invariant projection P_γ and positive constants K,ε. We now consider two cases: (i) $P_\gamma = \mathrm{id}$ and (ii) $P_\gamma = 0$:

Case (i). $P_\gamma = \mathrm{id}$, i.e., we have

$$\Phi_\alpha(t,\omega) \le K e^{(\gamma-\varepsilon)t} \quad \text{for all } t \ge 0\,. \tag{5.20}$$

Choose and fix $T > 0$ such that $e^{\frac{\varepsilon}{4}T} > K$. According to Proposition 5.1, there exists a measurable set $\mathcal{A} \subset \mathcal{F}_{-\infty}^{T}$ of positive measure such that

$$a_\alpha(\theta_s\omega) \in \big(-\sqrt{\varepsilon}/2, \sqrt{\varepsilon}/2\big) \quad \text{for all } \omega \in \mathcal{A} \text{ and } s \in [0,T]\,.$$

From (5.19), we derive that

$$|\Phi_\alpha(T,\omega)| \ge e^{T\left(\alpha - \frac{3\varepsilon}{4}\right)} > K e^{(\gamma-\varepsilon)T}\,,$$

which leads to a contradiction to (5.20).
Case (ii): $P_\gamma = 0$, i.e., we have

$$\Phi_\alpha(t,\omega) \ge \frac{1}{K} e^{(\gamma-\varepsilon)t} \quad \text{for all } t \ge 0\,,$$

which together with (5.19) implies that

$$\frac{\ln K + (\alpha-\gamma)t}{3} \ge \int_0^t a_\alpha(\theta_s\omega)^2\, \mathrm{d}s\,. \tag{5.21}$$

Choose and fix $T > 0$ such that

$$\frac{(T-1)^3}{3} > \frac{\ln K + (\alpha-\gamma)T}{3}\,.$$

Consider the following integral equation

$$x(t) = \int_0^t \big(\alpha x(s) - x(s)^3\big)\mathrm{d}s + \frac{t^4}{4} - \alpha\frac{t^2}{2} + t\,.$$

Clearly, the explicit solution of the above equation is $x(t) = t$. Due to the compactness, there exists $\varepsilon > 0$ such that for any $x(0) \in (-\varepsilon,\varepsilon)$ and $\omega(t)$ with $\sup_{t\in[0,T]} |\omega(t) - \frac{t^4}{4} + \alpha\frac{t^3}{3} - t| \le \varepsilon$ then the solution $x(t)$ of the following equation

$$x(t) = x(0) + \int_0^t (\alpha x(s) - x(s)^3)\, \mathrm{d}s + \omega(t)$$

satisfies that $\sup_{t\in[0,T]} |x(t) - t| \le 1$. According to Proposition 5.1, there exists a measurable set $\mathcal{A}_\varepsilon^- \subset \mathcal{F}_{-\infty}^0$ of positive measure such that $a_\alpha(\omega) \in (-\varepsilon,\varepsilon)$ for all $\omega \in \mathcal{A}$. Define $\mathcal{A}_+ \subset \mathcal{F}_0^T$ by

$$\mathcal{A}_\varepsilon^+ := \big\{\omega \in \mathcal{F}_0^T : \sup_{t\in[0,T]} |\omega(t) - t^4/4 + \alpha t^2/2 - t| \le \varepsilon\big\}\,.$$

Therefore, for all $\omega \in \mathcal{A}_\varepsilon^- \cap \mathcal{A}_\varepsilon^+$, we get

$$\sup_{t\in[0,T]} |a_\alpha(\theta_t\omega) - t| \le 1\,,$$

which implies that

$$\int_0^T a_\alpha(\theta_s\omega)^2\,\mathrm{d}s \ge \frac{(T-1)^3}{3} > \frac{\ln K + (\alpha-\gamma)T}{3}\,,$$

which leads to a contradiction to (5.21). The proof is complete.

We have seen in Theorem 5.4 that the bifurcation of (5.12) manifests itself also via finite-time Lyapunov exponents: Before the bifurcation, all finite-time Lyapunov exponents are negative, and after the bifurcation, one observes positive finite-time Lyapunov exponents with positive probability. This implies in particular that the set of all Lyapunov exponents observed almost surely within a *finite time* does not converge to the (asymptotic) Lyapunov exponents when time tends to infinity. The following theorem makes it precise that in contrast to asymptotic Lyapunov exponents, the dichotomy spectrum includes limits of the set of finite-time Lyapunov exponents.

Theorem 5.6 *Let* (θ, Φ) *be a linear random dynamical system on* $\mathbb{R}^d$ *with dichotomy spectrum* Σ. *Define the finite-time Lyapunov exponent*

$$\lambda(T,\omega,x) := \frac{1}{T}\ln\frac{\|\Phi(t,\omega)x\|}{\|x\|} \quad \textit{for all } T>0,\ \omega\in\Omega \textit{ and } x\in\mathbb{R}^d\setminus\{0\}\,.$$

Then

$$\lim_{T\to\infty} \operatorname*{ess\,sup}_{\omega\in\Omega} \sup_{x\in\mathbb{R}^d\setminus\{0\}} \lambda(T,\omega,x) = \sup\Sigma$$

provided that $\sup\Sigma < \infty$ *and*

$$\textit{and} \quad \lim_{T\to\infty} \operatorname*{ess\,inf}_{\omega\in\Omega} \inf_{x\in\mathbb{R}^d\setminus\{0\}} \lambda(T,\omega,x) = \inf\Sigma$$

provided that $\inf\Sigma > -\infty$.

Proof By definition of $\lambda(T,\omega,x)$, we get that for all $T, S \ge 0$

$$\begin{aligned}&(T+S)\operatorname*{ess\,sup}_{\omega\in\Omega}\sup_{x\in\mathbb{R}^d\setminus\{0\}}\lambda(T+S,\omega,x)\\ &\le T\operatorname*{ess\,sup}_{\omega\in\Omega}\sup_{x\in\mathbb{R}^d\setminus\{0\}}\lambda(T,\omega,x) + S\operatorname*{ess\,sup}_{\omega\in\Omega}\sup_{x\in\mathbb{R}^d\setminus\{0\}}\lambda(S,\omega,x).\end{aligned}$$

This implies that the function $[0, \infty) \cap \mathbb{T} \to \mathbb{R}, T \mapsto T \operatorname{ess\,sup}_{\omega \in \Omega} \sup_{x \in \mathbb{R}^d \setminus \{0\}} \lambda(T, \omega, x))$ is subadditive. We thus obtain

$$\lim_{T\to\infty} \operatorname*{ess\,sup}_{\omega\in\Omega} \sup_{x\in\mathbb{R}^d\setminus\{0\}} \lambda(T,\omega,x) = \limsup_{T\to\infty} \operatorname*{ess\,sup}_{\omega\in\Omega} \sup_{x\in\mathbb{R}^d\setminus\{0\}} \lambda(T,\omega,x).$$

We first prove that provided $\sup \Sigma < \infty$, we have

$$\gamma := \limsup_{T\to\infty} \operatorname*{ess\,sup}_{\omega\in\Omega} \sup_{x\in\mathbb{R}^d\setminus\{0\}} \lambda(T,\omega,x) = \sup \Sigma.$$

Since $\sup \Sigma < \infty$, it follows that there exists $K > 0$ such that

$$\|\Phi(t,\omega)x\| \leq K e^{t \sup \Sigma} \quad \text{for all } t \geq 0. \tag{5.22}$$

Assume first that $\gamma < \sup \Sigma$. This means that there exists a $t_0 > 0$ such that for all $t \geq t_0$ and for almost all $\omega \in \Omega$, we have $\|\Phi(t,\omega)x\| \leq e^{t/2(\gamma + \sup \Sigma)}$. Thus, together with (5.22), we obtain for all $t \geq 0$ and for almost all $\omega \in \Omega$ that

$$\|\Phi(t,\omega)x\| \leq \widehat{K} e^{t/2(\gamma+\sup\Sigma)}, \qquad \widehat{K} := \max\{1, K e^{t_0/2(\sup\Sigma - \gamma)}\}.$$

Hence, $\sup \Sigma \leq \gamma + \sup \Sigma$, which is a contradiction. Assume now that $\gamma > \sup \Sigma$. This means in particular that $\sup \Sigma < \infty$. Hence, there exists a $K > 0$ such that for almost all $\omega \in \Omega$, we have

$$\|\Phi(t,\omega)x\| \leq K e^{t/2(\gamma+\sup\Sigma)} \|x\| \quad \text{for all } x \in \mathbb{R}^d.$$

This leads to $\lambda(t, \omega, x) \leq (\gamma + \sup \Sigma)/2$ for all $x \in \mathbb{R}^d \setminus \{0\}$, and thus,

$$\gamma = \limsup_{T\to\infty} \operatorname*{ess\,sup}_{\omega\in\Omega} \sup_{x\in\mathbb{R}^d\setminus\{0\}} \lambda(T,\omega,x) \leq (\gamma + \sup\Sigma)/2,$$

which proves the first equality. Similarly, one can show that

$$\lim_{T\to\infty} \operatorname*{ess\,inf}_{\omega\in\Omega} \inf_{x\in\mathbb{R}^d\setminus\{0\}} \lambda(T,\omega,x) = \inf \Sigma$$

provided that $\inf \Sigma > -\infty$, which finishes the proof of this theorem.

In the following example, we construct explicitly a linear random dynamical system with $\sup \Sigma = \infty$ but

$$\lim_{T\to\infty} \operatorname*{ess\,sup}_{\omega\in\Omega} \sup_{x\in\mathbb{R}^d\setminus\{0\}} \lambda(T,\omega,x) < \infty.$$

An example of a linear random dynamical system with $\inf \Sigma = -\infty$ but

$$\lim_{T\to\infty} \operatorname*{ess\,inf}_{\omega\in\Omega} \inf_{x\in\mathbb{R}^d\setminus\{0\}} \lambda(T,\omega,x) > -\infty.$$

can be constructed analogously. This example shows the importance of the assumption $\sup \Sigma < \infty$ or $\inf \Sigma > -\infty$ in the above theorem.

Example 5.1 Similarly to Example 3.2, there exist infinitely many measurable sets $\{U_n\}_{n\in\mathbb{N}}$ of positive measure such that $U_n, \theta U_n, \theta^2 U_n$ for $n \in \mathbb{N}$ are pairwise disjoint. We define a random mapping $A : \Omega \to \mathbb{R}$ as follows:

$$A(\omega) = \begin{cases} \frac{1}{n} : \omega \in U_n \cup \theta^2 U_n\,, n \in \mathbb{N}\,, \\ n : \omega \in \theta U_n\,, n \in \mathbb{N}\,, \\ 1 : \text{otherwise.} \end{cases}$$

Let Φ denote the discrete-time random dynamical system generated by A. Since $\ln \|A(\cdot)\|$ is neither bounded from above nor from below, we get that $\Sigma(\Phi) = (-\infty, \infty)$. On the other hand, it is easy to see that for all $T \geq 2$, we get that

$$\operatorname*{ess\,sup}_{\omega\in\Omega} \ln|\Phi(T,\omega)| = 1,$$

which implies that

$$\lim_{T\to\infty} \operatorname*{ess\,sup}_{\omega\in\Omega} \frac{1}{T} \ln|\Phi(T,\omega)| = 0.$$

5.3 Random Hopf Bifurcation

Consider the two-dimensional stochastic differential equation

$$\begin{aligned} \mathrm{d}x &= (\alpha x - \beta y - (ax - by)(x^2 + y^2))\,\mathrm{d}t + \sigma\,\mathrm{d}W_t^1\,, \\ \mathrm{d}y &= (\alpha y + \beta x - (bx + ay)(x^2 + y^2))\,\mathrm{d}t + \sigma\,\mathrm{d}W_t^2\,, \end{aligned} \tag{5.23}$$

where $\sigma \geq 0$ represents the strength of the noise, $\alpha \in \mathbb{R}$ is a parameter equal to the real part of eigenvalues of the linearization of the vector field at $(0, 0)$, $b \in \mathbb{R}$ represents shear strength (amplitude-phase coupling parameter when writing the deterministic part of (5.23) in polar coordinates), $a > 0$, $\beta \in \mathbb{R}$, and W_t^1, W_t^2 denote independent one-dimensional Brownian motions.

In the absence of noise ($\sigma = 0$), the stochastic differential equation (5.23) is a normal form for the supercritical Hopf bifurcation: when $\alpha \leq 0$ the system has a globally attracting equilibrium at $(x, y) = (0, 0)$ which is exponentially stable until $\alpha = 0$ and, when $\alpha > 0$, the system has a limit cycle at $\{(x, y) \in \mathbb{R}^2 : x^2 + y^2 = \alpha/a\}$ which is globally attracting on $\mathbb{R}^2 \setminus \{0\}$.

In the presence of noise ($\sigma \neq 0$), by Sect. 5.1, the stochastic differential equation (5.23) generates a random dynamical system $\varphi : \mathbb{R}_0^+ \times \Omega \times \mathbb{R}^2 \to \mathbb{R}^2$ in the sense that $\Omega = C_0(\mathbb{R}, \mathbb{R}^2)$ be the space of all continuous functions $\omega : \mathbb{R} \to \mathbb{R}^2$ satisfying that $\omega(0) = 0$ and $\varphi(t, \omega, x)$ denotes the solution at time t of (5.23) which satisfies that $x(0) = x$. Furthermore, φ has a unique global random attractor A.

The rest of this section is devoted to studying the structure and the stability of the random attractor A. In Sect. 5.3.1, several sufficient conditions of parameters are provided for the collapse of random attractor, i.e., the attractor consists of only one point. This situation also means that the distance between any two trajectories converges to zero (this phenomena is called synchronization). Section 5.3.2 is devoted to bifurcation of random attractor in terms of qualitative change of finite-time behavior, change of the sign of finite-time Lyapunov spectrum, and dichotomy spectrum.

5.3.1 Synchronization

The following results concern the asymptotic behavior of trajectories, in particular their stability properties. This will give information about the structure of the random attractor A associated with the stochastic differential equation (5.23) which is also written as

$$\mathrm{d}Z_t = f(Z_t)\mathrm{d}t + \sigma \mathrm{d}W_t \,, \tag{5.24}$$

where $Z_t = (x_t, y_t)^\top$ and $W_t = (\mathrm{d}W_t^1, \mathrm{d}W_t^2)^\top$, and the function $f : \mathbb{R}^2 \to \mathbb{R}^2$ is defined by

$$f(Z) := \begin{pmatrix} \alpha & -\beta \\ \beta & \alpha \end{pmatrix} Z - (x^2 + y^2) \begin{pmatrix} a & -b \\ b & a \end{pmatrix} Z \,.$$

To analyze asymptotic stability, we study the linearization $\Phi(t, \omega, Z) := \mathrm{D}_x\varphi(t, \omega, Z)$. A direct computation yields that $\Phi(0, \omega, Z) = \mathrm{Id}$ and

$$\dot{\Phi}(t, \omega, Z) = \mathrm{D}f(\varphi(t, \omega, Z))\Phi(t, \omega, Z) \,. \tag{5.25}$$

It is easy to observe that Φ is a linear cocycle over the skew product flow $(\Theta_t)_{t \in \mathbb{R}_0^+}$ on $\Omega \times \mathbb{R}^2$, defined by

$$\Theta_t(\omega, Z) := (\theta_t\omega, \varphi(t, \omega, Z)) \,.$$

In fact, (Θ, Φ) is a linear random dynamical system, where the ergodic dynamical system $(\theta_t)_{t \in \mathbb{R}}$ is replaced by $(\Theta_t)_{t \in \mathbb{R}_0^+}$. We obtain an ergodic probability measure for the skew product flow $(\Theta_t)_{t\in\mathbb{R}_0^+}$ by using the fact that there exists a one-to-one correspondence between the stationary measure ρ for the Markov semigroup associated with (5.24) and a certain invariant measure of $(\Theta_t)_{t\in\mathbb{R}_0^+}$.

In more details, recall from [55] that the density of the unique stationary distribution ρ of (5.23) reads as

$$p(x, y) = K_{a,\alpha,\sigma} \exp\left(\frac{2\alpha(x^2 + y^2) - a(x^2 + y^2)^2}{2\sigma^2}\right) \quad \text{for all } (x, y) \in \mathbb{R}^2, \tag{5.26}$$

where $K_{a,\alpha,\sigma} > 0$ is the normalization constant and is given by

$$K_{a,\alpha,\sigma} = \frac{2\sqrt{2a}}{\sqrt{\pi}\sigma \operatorname{erfc}\left(-\frac{\alpha}{\sqrt{2a\sigma^2}}\right)}.$$

The stationary measure ρ gives rise to an invariant measure μ for $(\Theta_t)_{t\in\mathbb{R}_0^+}$ on $\Omega \times \mathbb{R}^2$ in the following sense: the push-forward limit

$$\mu_\omega := \lim_{t\to\infty} \varphi(t, \theta_{-t}\omega)\rho$$

exists for almost all $\omega \in \Omega$ and is an $\mathcal{F}_{-\infty}^0$-measurable random measure, i.e., $\omega \mapsto \mu_\omega(B)$ is $\mathcal{F}_{-\infty}^0$-measurable for any $B \in \mathcal{B}(\mathbb{R}^2)$. This defines a *Markov measure* μ on $(\Omega \times \mathbb{R}^2, \mathcal{F} \otimes \mathcal{B}(\mathbb{R}^2))$ via

$$\mu(C) := \int_\Omega \mu_\omega(C_\omega)\, \mathrm{d}\mathbb{P}(\omega) \quad \text{for all } C \in \mathcal{F} \otimes \mathcal{B}(\mathbb{R}^2),$$

where $C_\omega := \{Z \in \mathbb{R}^2 : (\omega, Z) \in C\}$. μ is invariant under $(\Theta_t)_{t\in\mathbb{R}_0^+}$ (see, e.g., [48]). Reversely, the stationary measure ρ is given by

$$\rho(B) = \int_\Omega \mu_\omega(B)\, \mathrm{d}\mathbb{P}(\omega) \quad \text{for all } B \in \mathcal{B}(\mathbb{R}^2). \tag{5.27}$$

The uniqueness of the stationary measure ρ with density $p(x, y)$ implies that the invariant measure μ is ergodic. We will show in the following Proposition 5.2 that the linear system Φ defined in (5.25) satisfies the integrability condition of multiplicative ergodic theorem ([6, Theorem 3.4.1] and Sect. 3.2.1), and therefore, we can apply this theorem to obtain the Lyapunov spectrum of the linear random dynamical system (Θ, Φ).

Proposition 5.2 *The following statements hold.*

(i) Let $\lambda^+ : \mathbb{R}^2 \to \mathbb{R}$ *be defined by*

$$\lambda^+(Z) := \max_{\|r\|=1} \langle Df(Z)r, r \rangle . \tag{5.28}$$

Then for almost all $\omega \in \Omega$ *and all* $Z \in \mathbb{R}^2$, *we have*

$$\|\Phi(t, \omega, Z)\| \le \exp\left(\int_0^t \lambda^+(\varphi(s, \omega, Z))\, ds\right) \quad \textit{for all } t \ge 0 , \tag{5.29}$$

and the linear random dynamical system Φ *satisfies the integrability condition of the multiplicative ergodic theorem.*

(ii) Let λ_Σ *be the sum of the two Lyapunov exponents of the linear random dynamical system* Φ. *Then* $\lambda_\Sigma < 0$ *and the disintegrations of the Markov measure* μ *are singular with respect to the Lebesgue measure on* $\mathbb{R}^2$.

Proof

(i) Let $v \in \mathbb{R}^2 \setminus \{0\}$ be arbitrary. By definition of Φ, we have

$$\begin{aligned}
\frac{d}{dt}\|\Phi(t, \omega, Z)v\|^2 &= 2\big\langle Df(\varphi(t, \omega, Z))\Phi(t, \omega, Z)v, \Phi(t, \omega, Z)v\big\rangle \\
&= 2\left\langle Df(\varphi(t, \omega, Z))\frac{\Phi(t, \omega, Z)v}{\|\Phi(t, \omega, Z)v\|}, \frac{\Phi(t, \omega, Z)v}{\|\Phi(t, \omega, Z)v\|}\right\rangle \\
&\quad \times \|\Phi(t, \omega, Z)v\|^2 \\
&\le 2\lambda^+(\varphi(t, \omega, Z))\|\Phi(t, \omega, Z)v\|^2.
\end{aligned}$$

This implies that

$$\|\Phi(t, \omega, Z)v\|^2 \le \|v\|^2 \exp\left(2\int_0^t \lambda^+(\varphi(s, \omega, Z))ds\right) .$$

Since v is arbitrary, (5.29) is proved. Using (5.29), we obtain that

$$\sup_{0\le t\le 1} \ln^+ \|\Phi(t, \omega, Z)\| \le \int_0^1 |\lambda^+(\varphi(s, \omega, Z))|\, ds ,$$

which implies that

$$\begin{aligned}
&\int_{\Omega\times\mathbb{R}^2} \sup_{0\le t\le 1} \ln^+ \|\Phi(t, \omega, Z)\|\, d\mu(\omega, Z) \\
&\quad \le \int_{\Omega\times\mathbb{R}^2} \int_0^1 |\lambda^+(\varphi(s, \omega, Z))|\, ds\, d\mu(\omega, Z)
\end{aligned}$$

$$= \int_0^1 \int_{\Omega \times \mathbb{R}^2} |\lambda^+(\varphi(s,\omega,Z))| \, \mathrm{d}\mu(\omega,Z) \, \mathrm{d}s$$

$$= \int_{\mathbb{R}^2} |\lambda^+(Z)| \, \mathrm{d}\rho(Z) , \tag{5.30}$$

where in the last equality, we use the fact that the skew product $\Theta_s(\omega, Z) = (\theta_s\omega, \varphi(s,\omega,Z))$ preserves the probability measure μ. By definition of λ^+ and the explicit form of $\mathrm{D}f$ given by

$$\mathrm{D}f(Z) = \begin{pmatrix} \alpha & -\beta \\ \beta & \alpha \end{pmatrix} - \begin{pmatrix} 3ax^2 + 2bxy + ay^2 & bx^2 + 2axy + 3by^2 \\ -3bx^2 - 2axy - by^2 & ax^2 - 2bxy + 3ay^2 \end{pmatrix},$$

it follows that

$$|\lambda^+(Z)| \le |\alpha| + 6(|a| + |b|)(x^2 + y^2) \quad \text{for all } Z = (x, y)^\top \in \mathbb{R}^2 .$$

Together with (5.30), this implies that

$$\int_{\Omega \times \mathbb{R}^2} \sup_{0 \le t \le 1} \ln^+ \|\Phi(t,\omega,Z)\| \, \mathrm{d}\mu(\omega,Z)$$
$$\le |\alpha| + 6(|a| + |b|) \int_{\mathbb{R}^2} (x^2 + y^2) p(x,y) \, \mathrm{d}x \, \mathrm{d}y ,$$

where $p(x, y)$ is given as in (5.26). Thus, the linear random dynamical system Φ satisfies the integrability condition of the multiplicative ergodic theorem.

(ii) Due to $\lambda_\Sigma = \lim_{t\to\infty} \frac{1}{t} \ln \det \Phi(t,\omega,Z)$, the sum of the two Lyapunov exponents of the linear random dynamical system generated by (5.25) reads as

$$\lambda_\Sigma = 2\alpha - 4a \int_{\mathbb{R}^2} (x^2 + y^2) p(x,y) \, \mathrm{d}x \, \mathrm{d}y .$$

Using the explicit formula for $p(x, y)$ from (5.26), we obtain that

$$\lambda_\Sigma = 2\alpha - 4a \frac{\int_{\mathbb{R}^2} (x^2 + y^2) \exp\left(\frac{2\alpha(x^2+y^2) - a(x^2+y^2)^2}{2\sigma^2}\right) \mathrm{d}x \, \mathrm{d}y}{\int_{\mathbb{R}^2} \exp\left(\frac{2\alpha(x^2+y^2) - a(x^2+y^2)^2}{2\sigma^2}\right) \mathrm{d}x \, \mathrm{d}y} .$$

Applying the change of variables $x = \sigma r \sin\varphi$, $y = \sigma r \cos\varphi$, the previous integral yields that

$$\lambda_\Sigma = 2\alpha - 4a\sigma^2 \frac{\int_0^\infty r^3 \exp\left(\frac{2\alpha r^2 - a\sigma^2 r^4}{2}\right) \mathrm{d}r}{\int_0^\infty r \exp\left(\frac{2\alpha r^2 - a\sigma^2 r^4}{2}\right) \mathrm{d}r}.$$

A further change of variable $r^2 \mapsto r$ gives that

$$\lambda_\Sigma = 2\alpha - 4a\sigma^2 \frac{\int_0^\infty r \exp\left(\frac{2\alpha r - a\sigma^2 r^2}{2}\right) \mathrm{d}r}{\int_0^\infty \exp\left(\frac{2\alpha r - a\sigma^2 r^2}{2}\right) \mathrm{d}r},$$

which proves that $\lambda_\Sigma < 0$ if $\alpha \leq 0$. We also show this for $\alpha > 0$ now. Using the change of variable $\sqrt{a}|\sigma|r - \frac{\alpha}{\sqrt{a}|\sigma|} \mapsto r$, we obtain that

$$\begin{aligned}\lambda_\Sigma &= -2\alpha - 4\sqrt{a}|\sigma| \frac{\int_{-\frac{\alpha}{|\sigma|\sqrt{a}}}^\infty r \exp\left(-\frac{r^2}{2}\right) \mathrm{d}r}{\int_{-\frac{\alpha}{|\sigma|\sqrt{a}}}^\infty \exp\left(-\frac{r^2}{2}\right) \mathrm{d}r} \\ &= -2\alpha - 4\sqrt{a}|\sigma| \frac{\exp\left(-\frac{\alpha^2}{2a\sigma^2}\right)}{\int_{-\frac{\alpha}{|\sigma|\sqrt{a}}}^\infty \exp\left(-\frac{r^2}{2}\right) \mathrm{d}r},\end{aligned}$$

which shows that $\lambda_\Sigma < 0$ for $\alpha > 0$. As a consequence, using [78, Proposition 1] and [20, Theorem 4.15], the disintegration of the Markov measure μ is singular with respect to ρ if $\lambda_\Sigma < 0$. The fact that ρ is equivalent to the Lebesgue measure finishes the proof of this proposition.

Therefore, we can apply Oseledets' multiplicative ergodic theorem [6, Theorem 3.4.1] to obtain the Lyapunov spectrum of the linear random dynamical system (Θ, Φ). In particular, the largest Lyapunov exponent λ_{top} is given by

$$\lambda_{\mathrm{top}} = \lim_{t\to\infty} \frac{1}{t} \ln \|\Phi(t, \omega, Z)\| \quad \text{for } \mu\text{-almost all } (\omega, Z) \in \Omega \times \mathbb{R}^2. \tag{5.31}$$

The first result of this subsection is the synchronization of the random dynamical system generated by (5.23) when its largest Lyapunov exponent λ_{top} is negative.

Theorem 5.7 (Existence of Random Equilibrium and Synchronization of Trajectories) *Suppose that $\lambda_{\mathrm{top}} < 0$. Then the random attractor A for the stochastic differential equation (5.23) is given by a random equilibrium, i.e., $A(\omega)$ is a singleton for almost all $\omega \in \Omega$. In addition, the stochastic differential equation (5.23) admits exponentially fast synchronization, i.e., for all $Z_1, Z_2 \in \mathbb{R}^2$, we have*

$$\limsup_{t\to\infty} \frac{1}{t} \ln \|\varphi(t, \omega, Z_1) - \varphi(t, \omega, Z_2)\| < 0 \quad \textit{for almost all } \omega \in \Omega.$$

Our proof of the preceding theorem consists of two ingredients. The first ingredient is a result from [63] that implies that the fibers of the random attractor are singletons. The second ingredient is the stable manifolds theorem, which we use to verify that this random attractor is also attractive in forward time.

We make use of the following sufficient conditions for the collapse of a random attractor [63, Theorem 2.14].

Theorem 5.8 (Collapse of the Random Attractor) *We assume that a random dynamical system* (θ, φ) *is*

(i) asymptotically stable *on a fixed non-empty open set* $U \subset \mathbb{R}^2$*, in the sense that there exists a sequence* $t_n \to \infty$ *such that*

$$\mathbb{P}\left(\omega \in \Omega : \lim_{n\to\infty} \operatorname{diam}(\varphi(t_n, \omega, U)) = 0\right) > 0\,.$$

(ii) swift transitive, *i.e., for all* $x, y \in \mathbb{R}^2$ *and* $r > 0$*, there exists a* $t > 0$ *such that*

$$\mathbb{P}\big(\omega \in \Omega : \varphi(t, \omega, B_r(x)) \subset B_{2r}(y)\big) > 0\,.$$

(iii) contracting on large sets, *i.e., for all* $R > 0$*, there exist* $y \in \mathbb{R}^2$ *and* $t > 0$ *such that*

$$\mathbb{P}\left(\omega \in \Omega : \operatorname{diam}(\varphi(t, \omega, B_R(y))) \le \tfrac{R}{4}\right) > 0\,.$$

Suppose further that (θ, φ) *has a random attractor* A *with* $\mathcal{F}^0_{-\infty}$*-measurable fibers. Then* $A(\omega)$ *is a singleton* $\mathbb{P}$*-almost surely.*

We use this result for the following proposition.

Proposition 5.3 *Suppose that the largest Lyapunov exponent* λ_{top} *of the random dynamical system generated by (5.23) is negative. Then the fibers of the random attractor are singletons, given by* $\mathcal{F}^0_{-\infty}$*-measurable map* $A : \Omega \to \mathbb{R}^2$*. Furthermore, the following statements hold:*

(i) *A is a random equilibrium of* φ*, i.e.,*

$$\varphi(t, \omega, A(\omega)) = A(\theta_t\omega) \quad \textit{for all } t \ge 0 \textit{ and almost all } \omega \in \Omega\,.$$

(ii) *The random equilibrium is distributed according to the stationary density* $(x, y) \mapsto p(x, y)$*; see (5.26). More precisely,*

$$\mathbb{P}\big(\{\omega \in \Omega : A(\omega) \in C\}\big) = \int_C p(x, y)\,\mathrm{d}x\,\mathrm{d}y \quad \textit{for all } C \in \mathcal{B}(\mathbb{R}^2)\,.$$

(iii) *The largest Lyapunov exponent of the linearization along the random equilibrium a,*

$$\dot{\xi} = \mathrm{D}f(A(\theta_t\omega))\xi\,,$$

is equal to λ_{top} *as given in* (5.31).

Proof In the first part of the proof, we show that the random dynamical system φ generated by (5.23) fulfils the assumptions (i), (ii), and (iii) of Theorem 5.8. Note that (i) follows from the negativity of the largest Lyapunov exponent (see [63, Lemma 4.1 and Corollary 4.4]), and swift transitivity holds for our system according to [63, Proposition 4.9]. Hence, it remains to show contraction on large sets for φ. By definition of f, we have that

$$\begin{aligned}\big\langle f(x)-f(y), x-y\big\rangle &\le \left(\alpha - a\tfrac{1}{2}(\|x\|^2+\|y\|^2)\right)\|x-y\|^2\\ &\quad + b(x_1y_2-y_1x_2)\left(2\langle x-y, y\rangle + \|x-y\|^2\right).\end{aligned}$$

Fix $r>0$, and consider $B_r(z)$, where $z=(R,0)$ for some $R>0$ to be chosen large enough. For any $x, y \in B_r(z)$, observe that

$$(x_1y_2-y_1x_2)\langle x-y, y\rangle \le r\|y\|\|x-y\|^2 + r^2\|x-y\|^2$$

and

$$(x_1y_2-y_1x_2)\|x-y\|^2 \le 2\|x-y\|^2\|y\|\|x-y\|\,.$$

This implies that for all $x, y \subset B_r(z)$,

$$\begin{aligned}\langle f(x)-f(y), x-y\rangle &\le \|x-y\|^2\left(\alpha - a\tfrac{1}{2}(\|x\|^2+\|y\|^2)\right.\\ &\qquad \left. +2b(r\|y\| + r^2 + 2\|y\|r)\right)\\ &< K\|x-y\|^2\end{aligned}$$

for some $K<0$ if R is big enough (due to the quadratic terms, K has negative sign). This property is called monotonicity on large sets, which implies contraction on large sets due to [63, Proposition 3.10].

We now prove the statements (i), (ii), and (iii) of the proposition.

(i) This follows immediately from the definition of a random attractor (see Definition 5.1).
(ii) Note that $\omega \mapsto A(\omega)$ is measurable with respect to $\mathcal{F}^0_{-\infty}$, and thus, $\mu_\omega := \delta_{A(\omega)}$ defines a Markov measure. The invariance of μ_ω follows directly from (i). Hence, $\{\mu_\omega\}_{\omega\in\Omega}$ is the disintegration of the ergodic invariant measure μ associated with the ergodic stationary measure ρ, and we obtain from (5.27) that for all $C \in \mathcal{B}(\mathbb{R}^2)$

$$\begin{aligned}\mathbb{P}(\{\omega \in \Omega : A(\omega) \in C\}) &= \int_\Omega \delta_{A(\omega)}(C) \, \mathrm{d}\mathbb{P}(\omega) = \int_\Omega \mu_\omega(C) \, \mathrm{d}\mathbb{P}(\omega) \\ &= \rho(U) = \int_C p(x, y) \, \mathrm{d}x \, \mathrm{d}y \,.\end{aligned}$$

(iii) According to the multiplicative ergodic theorem, the existence of the Lyapunov spectrum holds for a set $M \subset \Omega \times \mathbb{R}^2$ of full μ-measure. We observe that the set

$$D = \bigcup_{\omega \in \Omega} \{(\omega, A(\omega))\}$$

has full μ-measure, since

$$\mu(D) = \int_\Omega \mu_\omega(\{A(\omega)\}) \, \mathrm{d}\mathbb{P}(\omega) = \int_\Omega \delta_{A(\omega)}(\{A(\omega)\}) \, \mathrm{d}\mathbb{P}(\omega) = 1 \,.$$

Hence, $\mu(M \cap D) = 1$. Since the Oseledets space associated with the second Lyapunov exponent has zero Lebegue measure for any $(\omega, x) \in M \cap D$, the claim follows.

We are now in a position to prove Theorem 5.7.

Proof of Theorem 5.7 The existence of the attracting random equilibrium $A : \Omega \to \mathbb{R}^2$ has been shown in Proposition 5.3. Define $\psi : \mathbb{R}_0^+ \times \Omega \times \mathbb{R}^2 \to \mathbb{R}^2$ by

$$\psi(t, \omega, x) = \varphi(t, \omega, A(\omega) + x) - \varphi(t, \omega, A(\omega)).$$

Obviously, $\psi(t, \omega, 0) = 0$ and $\psi(t, \omega, x)$ are the solution of the random differential equation

$$\dot{\xi} = \mathrm{D}f(A(\theta_t\omega))\xi + R(t, \omega, \xi), \tag{5.32}$$

where

$$R(t, \omega, \xi) := f(A(\theta_t\omega) + \xi) - f(A(\theta_t\omega)) - Df(A(\theta_t\omega))\xi \,.$$

Note that for $R \equiv 0$, the largest Lyapunov exponent of the homogeneous equation (5.32) is negative. Using the stable manifold theorem [6, Theorems 7.5.5 and 7.5.16], there exists $r(\omega) > 0$ such that for almost all $\omega \in \Omega$ and $x \in B_{r(\omega)}(0)$, one has

$$\lim_{t\to\infty} e^{-\frac{\lambda_{\mathrm{top}}}{2}t} \|\psi(t, \omega, x)\| = \lim_{t\to\infty} e^{-\frac{\lambda_{\mathrm{top}}}{2}t} \|\varphi(t, \omega, x + A(\omega)) - A(\theta_t\omega)\| = 0 \,. \tag{5.33}$$

Choose and fix an arbitrary initial value $x \in \mathbb{R}^2$, and define

$$V := \left\{ \omega \in \Omega : \limsup_{t\to\infty} e^{-\frac{\lambda_{\rm top}}{2} t} \|\varphi(t,\omega,x) - A(\theta_t\omega)\| = 0 \right\}.$$

It remains to show that $\mathbb{P}(V) = 1$. For each $n \in \mathbb{N}$, we define

$$\Omega_n := \left\{ \omega \in \Omega : \varphi(t, \theta_{-t}\omega, x) \in B_{r(\omega)}(A(\omega)) \text{ for all } t \geq n \right\}.$$

Note that $(\Omega_n)_{n\in\mathbb{N}}$ is an increasing sequence of measurable sets. By virtue of Proposition 5.3, the random equilibrium a is the random attractor of φ, which implies $\lim_{n\to\infty} \mathbb{P}(\Omega_n) = 1$. From the definition of Ω_n, we derive that $\varphi(n, \theta_{-n}\omega, x) \in B_{r(\omega)}(A(\omega))$ for all $\omega \in \Omega_n$. Together with (5.33), this implies that for all $\omega \in \Omega_n$, one has

$$\begin{aligned} 0 &= \limsup_{t\to\infty} e^{-\frac{\lambda_{\rm top}}{2} t} \|\varphi(t,\omega,\varphi(n,\theta_{-n}\omega,x)) - A(\theta_t\omega)\| \\ &= \limsup_{t\to\infty} e^{-\frac{\lambda_{\rm top}}{2} t} \|\varphi(t+n,\theta_{-n}\omega,x) - A(\theta_t\omega)\|. \end{aligned}$$

Consequently, $\theta_{-n}\Omega_n \subset V$, and thus, $\mathbb{P}(V) = 1$, which finishes the proof.

The last result of this subsection is to determine the region of parameters for which $\lambda_{\rm top} < 0$. In [55], analytical results are obtained that show that $\lambda_{\rm top}$ is negative in certain regions of the parameters space, in particular when shear is small. The following theorem extends this result to a larger region in the parameter space.

Theorem 5.9 (Small Shear Implies Synchronization) *For each a, α, β, σ, let*

$$\kappa := a\sqrt{\frac{\pi K_{a,\alpha,\sigma}\sigma^2}{\alpha + \pi K_{a,\alpha,\sigma}\sigma^2}\left(\frac{\pi K_{a,\alpha,\sigma}\sigma^2}{\alpha + \pi K_{a,\alpha,\sigma}\sigma^2} + 2\right)}.$$

Then the largest Lyapunov exponent $\lambda_{\rm top}$ is negative if $|b| \leq \kappa$.

Remark 5.3

(i) Note that $K_{a,0,\sigma} = \frac{2\sqrt{2a}}{\sqrt{\pi}\sigma}$, and Theorem 5.9 then implies that $\lambda_{\rm top} < 0$ provided that $|b| < \sqrt{3}a$ and α is sufficiently small. This special case is considered in [55, Proposition 4.1].

(ii) For fixed a and α, we have

$$\lim_{\sigma\to\infty} \pi K_{a,\alpha,\sigma}\sigma^2 = \lim_{\sigma\to\infty} \frac{2\sqrt{2\pi a}\sigma}{\operatorname{erfc}\left(-\frac{\alpha}{\sqrt{2a\sigma^2}}\right)} = \infty.$$

The main ingredient for the proof of Theorem 5.9 is the inequality in Proposition 5.2(i). We first need the following estimate on the function λ^+ defined as in (5.28).

Lemma 5.3 *For any* $Z = (x, y)^\top \in \mathbb{R}^2$, *we have*

$$\lambda^+(Z) \le \alpha + \left(\sqrt{a^2+b^2} - 2a\right)(x^2+y^2)\,,$$

and equality holds if and only if $xy = 0$.

Proof Using the following explicit form of $Df(Z)$,

$$Df(Z) = \begin{pmatrix} \alpha - ay^2 - 3ax^2 - 2byx & -\beta - 2axy - bx^2 - 3by^2 \\ \beta - 2axy + by^2 + 3bx^2 & \alpha - ax^2 - 3ay^2 + 2byx \end{pmatrix},$$

we obtain for any $r \in \mathbb{R}^2$ with $\|r\| = 1$ that

$$\begin{aligned} \langle Df(x,y)r, r\rangle &= r_1^2(\alpha - ay^2 - 3ax^2) + r_1r_2(-\beta - 2axy) + r_1r_2(\beta - 2axy) \\ &\quad + r_2^2(\alpha - ax^2 - 3ay^2) \\ &\quad - 2byxr_1^2 + 2byxr_2^2 + r_1r_2(2bx^2 - 2by^2) \\ &= \alpha - a(x^2+y^2) + 2b(r_1r_2x^2 - r_1r_2y^2 + (r_2^2 - r_1^2)xy) \\ &\quad - 2a(r_1x + r_2y)^2\,. \end{aligned}$$

Since $r_1^2 + r_2^2 = 1$, it is possible to write that $r_1 = \sin\varphi$ and $r_2 = \cos\varphi$ for some $\varphi \in [0, 2\pi)$. Thus, a simple calculation yields that

$$\begin{aligned} &\langle Df(x,y)r, r\rangle \\ &= \alpha - 2a(x^2+y^2) + (ax^2 + bxy - ay^2)\cos 2\varphi + (bx^2 - 2axy - by^2)\sin 2\varphi \\ &\le \alpha - 2a(x^2+y^2) + \sqrt{(ax^2 + 2bxy - ay^2)^2 + (bx^2 - 2axy - by^2)^2} \\ &= \alpha - 2a(x^2+y^2) + \sqrt{(a^2+b^2)(x^2-y^2)^2 + 4b^2x^2y^2} \\ &\le \alpha - 2a(x^2+y^2) + \sqrt{a^2+b^2}(x^2+y^2)\,, \end{aligned}$$

which completes the proof.

Proof of Theorem 5.9 From inequality (5.29), we derive that

$$\lambda_{\text{top}} = \limsup_{t\to\infty} \frac{1}{t} \ln \|\Phi(t,\omega,s)\| \le \lim_{t\to\infty} \frac{1}{t}\int_0^t \lambda^+(\varphi(s,\omega,x))\,\mathrm{d}s\,.$$

Note that the skew product flow $\Theta_s(\omega, Z) = (\theta_s\omega, \varphi(s,\omega,Z))$ preserves the probability measure μ, and λ^+ is integrable. By using Birkhoff's ergodic theorem [122, Theorem 1.14], we obtain that

$$\lambda_{\text{top}} \leq \int_{\mathbb{R}^2} \lambda^+(x, y) p(x, y) \, dx \, dy ,$$

where the density function p is as in (5.26). Thus, by virtue of Lemma 5.3, we arrive at

$$\lambda_{\text{top}} < \alpha + \left(\sqrt{a^2 + b^2} - 2a\right) \int_{\mathbb{R}^2} (x^2 + y^2) p(x, y) \, dx \, dy .$$

Inserting the explicit form of the density function p in the preceding inequality gives that

$$\begin{aligned} \lambda_{\text{top}} &< \alpha + (\sqrt{a^2 + b^2} - 2a) K \int_{\mathbb{R}^2} (x^2 + y^2) \\ &\quad \times \exp\left(\frac{2\alpha(x^2 + y^2) - a(x^2 + y^2)^2}{2\sigma^2}\right) dx \, dy , \end{aligned} \tag{5.34}$$

with the normalization constant $K = \frac{2\sqrt{2a}}{\sqrt{\pi}\sigma \operatorname{erfc}\left(-\alpha/\sqrt{2a\sigma^2}\right)}$. Using polar coordinates, we obtain that

$$\begin{aligned} &K \int_{\mathbb{R}^2} \left(\alpha - a(x^2 + y^2) \exp\left(\frac{2\alpha(x^2 + y^2) - a(x^2 + y^2)^2}{2\sigma^2}\right)\right) dx \, dy \\ &= 2\pi K \int_0^\infty (\alpha - ar^2) r \exp\left(\frac{2\alpha r^2 - ar^4}{2\sigma^2}\right) dr \\ &= -\pi\sigma^2 K . \end{aligned}$$

This implies

$$\int_{\mathbb{R}^2} (x^2 + y^2) \exp\left(\frac{2\alpha(x^2 + y^2) - a(x^2 + y^2)^2}{2\sigma^2}\right) dx \, dy = \frac{\alpha}{Ka} + \frac{\pi\sigma^2}{a} ,$$

which together with (5.34) implies that

$$\lambda_{\text{top}} < \alpha + \left(\sqrt{a^2 + b^2} - 2a\right) \left(\frac{\alpha}{a} + \frac{\pi K \sigma^2}{a}\right) .$$

Consequently,

$$\lambda_{\text{top}} < -\pi K \sigma^2 + \left(\sqrt{1 + \frac{b^2}{a^2}} - 1\right) (\alpha + \pi K \sigma^2) .$$

Note that by definition of K, it is easy to see that $\alpha + \pi K\sigma^2 > 0$. Therefore, for all $|b| \leq \kappa$, we have

$$\lambda_{\text{top}} < -\pi K\sigma^2 + \left(\sqrt{1 + \frac{\kappa^2}{a^2}} - 1\right)(\alpha + \pi K\sigma^2) = 0\,,$$

which completes the proof of this theorem.

5.3.2 Random Hopf Bifurcation

We analyze random bifurcations for the stochastic differential equation (5.23) in this subsection, which captures qualitative changes in the the asymptotic as well as the finite-time behavior. For this purpose, we first need the following preparatory proposition.

Proposition 5.4 *Consider* (5.23) *such that* $|b| \leq \kappa$. *Then for any* $y \in \mathbb{R}^2$, $\epsilon > 0$, *and* $T \geq 0$, *there exists a set* $E \in \mathcal{F}_{-\infty}^T$ *with* $\mathbb{P}(E) > 0$ *such that*

$$A(\theta_s\omega) \in B_\epsilon(y) \quad \textit{for all } s \in [0, T] \textit{ and } \omega \in E\,,$$

where $\{A(\omega)\}$ *is the unique random equilibrium for* (5.23) *from Proposition 5.3.*

Proof Let $\epsilon > 0$ and $T \geq 0$. Since $\Omega = \bigcup_{x\in\mathbb{Q}^2}\{\omega \in \Omega : A(\omega) \in B_{\epsilon/4}(x)\}$, there exists an $x \in \mathbb{R}^2$ such that

$$A_0 := \{\omega \in \Omega : A(\omega) \in B_{\epsilon/4}(x)\}$$

has positive measure. From [63, Proposition 3.10], we know that there exists $t_0 > 0$ such that

$$B_0 := \left\{\omega \in \Omega : \varphi(t_0, \omega, x') \in B_{\epsilon/2}(y) \text{ for all } x' \in B_{\epsilon/4}(x)\right\}$$

has positive measure. Since θ is measure preserving, the two sets

$$\begin{aligned}
A_1 &:= \theta_{t_0} A_0 = \{\omega \in \Omega : A(\theta_{-t_0}\omega) \in B_{\epsilon/4}(x)\}\,,\\
B_1 &:= \theta_{t_0} B_0 = \left\{\omega \in \Omega : \varphi(t_0, \theta_{-t_0}\omega, x') \in B_{\epsilon/2}(y) \text{ for all } x' \in B_{\epsilon/4}(x)\right\}
\end{aligned}$$

have positive measure. Due to the Markov property of the random dynamical system, we observe that B_1 and A_1 are independent, and hence, $\mathbb{P}(B_1 \cap A_1) > 0$. Thus, the set

$$E_0 = \{\omega \in \Omega : A(\omega) \in B_{\epsilon/2}(y)\} \supset A_1 \cap B_1$$

has positive measure and clearly lies in $\mathcal{F}_{-\infty}^{0}$. Fix $\omega \in E_0$. Similarly to the proof of [63, Proposition 3.10], define

$$h(t) := -\frac{tf(A(\omega))}{\sigma} \quad \text{for all } t \in [0, T],$$

where f denotes the vector field of the drift in (5.24). We write $\varphi(t, g, z)$, $t \in [0, T]$, for the solution of (5.24) with initial condition z and path $g \in C_0^T := \{\bar{g} \in C([0, T], \mathbb{R}^2) : \bar{g}(0) = 0\}$. We can infer that $\varphi(t, h, A(\omega)) = A(\omega)$ for all $t \in [0, T]$. Recall that the map $g \mapsto \varphi(\cdot, g, z)$ is continuous from C_0^T to $C([0, T], \mathbb{R}^2)$ with respect to the supremum norm $\|\cdot\|_\infty$. Hence, there is a $\delta > 0$ such that for all $g \in C_\delta := \{\bar{g} \in C_0^T : \|\bar{g} - h\| \leq \delta\}$, we have

$$\|\varphi(t, g, A(\omega)) - \varphi(t, h, A(\omega))\| < \epsilon/2 \quad \text{for all } t \in [0, T].$$

Since the set $E_+ := \{\omega : \omega|_{[0,T]} \in C_\delta\}$ has positive measure and is independent of E_0, the set $E = E_0 \cap E_+ \in \mathcal{F}_{-\infty}^{T}$ has positive measure and satisfies

$$A(\theta_t\omega) \in B_\epsilon(y) \quad \text{for all } t \in [0, T] \text{ and } \omega \in E,$$

by the above construction.

In this subsection, we consider the stochastic differential equation (5.23) with small enough shear such there exists a random equilibrium for α close to zero. In the first main result, we prove in Theorem 5.10 that the random equilibrium $A : \Omega \to \mathbb{R}^2$ loses uniform attractivity at the deterministic bifurcation point $\alpha = 0$. On the other hand, we will observe a loss of hyperbolicity at the bifurcation point in the dichotomy spectrum associated with the random equilibrium. Moreover, we can show that $A : \Omega \to \mathbb{R}^2$ is finite-time attracting before, but not after the bifurcation point, indicated by a transition from zero to positive probability of positive finite-time Lyapunov exponents.

We call the random attractor A *locally uniformly attracting* if there exists a $\delta > 0$ such that

$$\lim_{t\to\infty} \sup_{x\in B_\delta(0)} \operatorname{ess\,sup}_{\omega\in\Omega} \|\varphi(t, \omega, A(\omega) + x) - A(\theta_t\omega)\| = 0.$$

We call it *globally uniformly attracting* if the above holds for any $\delta > 0$.

Theorem 5.10 *Consider the stochastic differential equation* (5.23) *such that there is a unique attracting random equilibrium $A : \Omega \to \mathbb{R}^2$ (see Proposition 5.3 and Theorem C). Then for $\alpha < 0$ and $|b| \leq a$, the random attractor $A : \Omega \to \mathbb{R}^2$ is globally uniformly attracting. Furthermore, for all pairs of initial conditions $U, V \in \mathbb{R}^2$, we have*

$$\|\varphi(t, \omega, U) - \varphi(t, \omega, V)\| \leq e^{\alpha t}\|U - V\| \quad \textit{for all } t \geq 0.$$

For $\alpha > 0$, the random attractor $A : \Omega \to \mathbb{R}^2$ is not even locally uniformly attracting.

Proof Fix $\alpha < 0$, and choose arbitrary $U, V \in \mathbb{R}^2$, $\omega \in \Omega$. Define

$$\begin{pmatrix} x_t \\ y_t \end{pmatrix} := \varphi(t, \omega, U) \quad \text{and} \quad \begin{pmatrix} \widehat{x}_t \\ \widehat{y}_t \end{pmatrix} := \varphi(t, \omega, V) .$$

From (5.23), we derive that

$$\frac{\mathrm{d}}{\mathrm{d}t} \begin{pmatrix} x_t - \widehat{x}_t \\ y_t - \widehat{y}_t \end{pmatrix} = \begin{pmatrix} \alpha & -\beta \\ \beta & \alpha \end{pmatrix} \begin{pmatrix} x_t - \widehat{x}_t \\ y_t - \widehat{y}_t \end{pmatrix} - \\ (x_t^2 + y_t^2) \begin{pmatrix} a & b \\ -b & a \end{pmatrix} \begin{pmatrix} x_t \\ y_t \end{pmatrix} + (\widehat{x}_t^2 + \widehat{y}_t^2) \begin{pmatrix} a & b \\ -b & a \end{pmatrix} \begin{pmatrix} \widehat{x}_t \\ \widehat{y}_t \end{pmatrix} .$$

Therefore,

$$\begin{aligned} \frac{1}{2} \frac{\mathrm{d}}{\mathrm{d}t} \left\| \begin{pmatrix} x_t - \widehat{x}_t \\ y_t - \widehat{y}_t \end{pmatrix} \right\|^2 &= (x_t - \widehat{x}_t) \frac{\mathrm{d}}{\mathrm{d}t} (x_t - \widehat{x}_t) + (y_t - \widehat{y}_t) \frac{\mathrm{d}}{\mathrm{d}t} (y_t - \widehat{y}_t) \\ &= \alpha \left\| \begin{pmatrix} x_t - \widehat{x}_t \\ y_t - \widehat{y}_t \end{pmatrix} \right\|^2 - R(x_t, \widehat{x}_t, y_t, \widehat{y}_t) , \end{aligned}$$

where

$$R(x_t, y_t, \widehat{x}_t, \widehat{y}_t) := a \left(r_t^2 + \widehat{r}_t^2 - (x_t \widehat{x}_t + y_t \widehat{y}_t)(r_t + \widehat{r}_t) \right) + b(x_t \widehat{y}_t - \widehat{x}_t y_t)(r_t - \widehat{r}_t)$$

with $r_t := x_t^2 + y_t^2$ and $\widehat{r}_t := \widehat{x}_t^2 + \widehat{y}_t^2$. To show global uniform attractivity, it is sufficient to establish that $R(x_t, y_t, \widehat{x}_t, \widehat{y}_t) \geq 0$. From the inequality $(|xy| + |uv|)^2 \leq (x^2 + u^2)(y^2 + v^2)$, we derive that

$$\begin{aligned} &|(x_t \widehat{x}_t + y_t \widehat{y}_t)(r_t + \widehat{r}_t)| + |(x_t \widehat{y}_t - \widehat{x}_t y_t)(r_t - \widehat{r}_t)| \\ &\leq \sqrt{(x_t \widehat{x}_t + y_t \widehat{y}_t)^2 + (x_t \widehat{y}_t - \widehat{x}_t y_t)^2} \sqrt{(r_t + \widehat{r}_t)^2 + (r_t - \widehat{r}_t)^2} \\ &= \sqrt{2 r_t \widehat{r}_t (r_t^2 + \widehat{r}_t^2)} \\ &\leq r_t^2 + \widehat{r}_t^2 . \end{aligned}$$

Together with the fact that $|b| \leq a$, this implies that $R(x_t, y_t, \widehat{x}_t, \widehat{y}_t) \geq 0$, which establishes global uniform attractivity for $\alpha < 0$.

We assume now that $\alpha > 0$. Suppose to the contrary that there exists $\delta > 0$ such that

$$\lim_{t\to\infty} \sup\nolimits_{x\in B_\delta(0)} \operatorname{ess\,sup}_{\omega\in\Omega} \|\varphi(t,\omega,A(\omega)+x)-A(\theta_t\omega)\| = 0\,.$$

This implies that there exists an $N \in \mathbb{N}$ such that for all $t > N$, we have

$$\sup\nolimits_{x\in B_\delta(0)} \operatorname{ess\,sup}_{\omega\in\Omega} \|\varphi(t,\omega,A(\omega)+x)-A(\theta_t\omega)\| < \tfrac{1}{4}\sqrt{\tfrac{\alpha}{a}}\,.$$

Due to Proposition 5.4, there exists a positive measure set $E_0 \in \mathcal{F}_{-\infty}^{0}$ such that $A(\omega) \in B_{\delta/4}(0)$ for all $\omega \in E_0$. Let $\varphi(\cdot, x_0)$ denote the solution of the deterministic equation (5.23) for $\sigma = 0$ with initial condition $x(0) = x_0$. Then there exists a $T > N$ such that

$$\|\varphi(T,(\pm\tfrac{1}{4}\delta,0))\| > \tfrac{1}{2}\sqrt{\tfrac{\alpha}{a}}\,,$$

and at the same time

$$\|\varphi(T,(\tfrac{1}{4}\delta,0))-\varphi(T,(-\tfrac{1}{4}\delta,0))\| > \sqrt{\tfrac{\alpha}{a}}\,.$$

Recall from the proof of Proposition 5.4 that $\omega \mapsto \varphi(\cdot,\omega,x)$ is continuous from C_0^T to $C([0,T],\mathbb{R}^2)$ with respect to the supremum norm. This implies that there exists an $\epsilon > 0$ such that for all $\omega \in E_\epsilon = \{\omega\in\Omega : \sup_{t\in[0,T]}\|\omega(t)\| < \epsilon\} \in \mathcal{F}_0^T$, we obtain

$$\varphi\left(T,\left(\frac{1}{4}\delta,0\right)\right)-\varphi\left(T,\omega,\left(\frac{1}{4}\delta,0\right)\right)\| < \frac{1}{4}\sqrt{\frac{\alpha}{a}},$$
$$\|\varphi\left(T,\left(-\frac{1}{4}\delta,0\right)\right)-\varphi\left(T,\omega,\left(-\frac{1}{4}\delta,0\right)\right)\| < \frac{1}{4}\sqrt{\frac{\alpha}{a}}.$$

This implies that

$$\|\varphi(T,\omega,(\tfrac{1}{4}\delta,0))-\varphi(T,\omega,(-\tfrac{1}{4}\delta,0))\| > \tfrac{1}{2}\sqrt{\tfrac{\alpha}{a}}\,.$$

Since E_ϵ and E_0 are independent sets of positive measure, we get that $\mathbb{P}(E) > 0$ where $E = E_\epsilon \cap E_0$. However, for all $\omega \in E$, we conclude

$$\begin{aligned}&\sup_{x\in B_\delta(0)} \|\varphi(t,\omega,A(\omega)+x)-A(\theta_t\omega)\|\\ &\geq \max\Big\{\big\|\varphi(t,\omega,(\tfrac{1}{4}\delta,0))-A(\theta_t\omega)\big\|\,,\ \big\|\varphi(t,\omega,(-\tfrac{1}{4}\delta,0))-A(\theta_t\omega)\big\|\Big\} > \tfrac{1}{4}\sqrt{\tfrac{\alpha}{a}}\,,\end{aligned}$$

which contradicts our assumption.

In the second main result, we show now that this loss of uniform attractivity at the deterministic bifurcation point is associated with a change of sign in the dichotomy spectrum.

Theorem 5.11 *Consider the stochastic differential equation* (5.23) *such that there exists a unique attracting random equilibrium* $A : \Omega \to \mathbb{R}^2$ *(see Proposition 5.3 and Theorem C). Let* $\Phi(t, \omega) := \mathrm{D}\varphi(t, \omega, A(\omega))$ *denote the linearized random system along the random equilibrium. Then for* $|b| < a$ *and* $\alpha \in \mathbb{R}$ *small enough such that the random equilibrium* $A : \Omega \to \mathbb{R}^2$ *exists, the dichotomy spectrum* Σ *of* Φ *is given by*

$$\Sigma = [-\infty, \alpha] .$$

Proof Recall from Proposition 5.2 that we have

$$\|\mathrm{D}\varphi(t, \omega, x)\| \le \exp\left(\int_0^t \lambda^+(\varphi(s, \omega, x))\mathrm{d}s\right) .$$

Since Lemma 5.3 implies that $\lambda^+(x) \le \alpha - (a - |b|)\|x\|^2$, we have

$$\|\Phi(t, \omega)\| \le \exp\left(\int_0^t \left(\alpha - (a - |b|)\|A(\theta_s\omega))\|^2\right) \mathrm{d}s\right) . \tag{5.35}$$

Similarly, with $\lambda^-(x) := \min_{\|r\|=1}\langle Df(x)r, r\rangle$, we have

$$\|D\varphi(t, \omega, x)\| \ge \exp\left(\int_0^t \lambda^-(\varphi(s, \omega, x))\mathrm{d}s\right) .$$

It is easy to see that $\lambda^-(x) \ge \alpha - 4a\|x\|^2$, which implies

$$\|\Phi(t, \omega)\| \ge \exp\left(\int_0^t (\alpha - 4a\|A(\theta_s\omega))\|^2)\mathrm{d}s\right) . \tag{5.36}$$

From (5.35), we can deduce immediately that for almost all $\omega \in \Omega$, we have

$$\|\Phi(t, \omega)\| \le e^{\alpha|t|} \quad \text{for all } t \in \mathbb{R} .$$

This implies that $\Sigma \subset (-\infty, \alpha]$.

We now show that $(-\infty, \alpha] \subset \Sigma$. Choose $\gamma \in (-\infty, \alpha]$, and suppose to the contrary that Φ admits an exponential dichotomy with growth rate γ with an invariant projector P_γ and constants $K, \epsilon > 0$. We consider the following three cases (note that the rank of the invariant projector rk P_γ does not depend on ω, see Proposition 3.1):

(i) $P_\gamma \equiv \mathrm{id}$. This means that for almost all $\omega \in \Omega$,

$$\|\Phi(t, \omega)\| \leq K e^{(\gamma-\epsilon)t} \quad \text{for all } t \geq 0 .$$

Fix $T > 0$ such that $e^{\frac{1}{5}\epsilon T} > K$. According to Proposition 5.4, there exists a positive measure set E such that for all $\omega \in E$ and $s \in [0, T]$, we have $A(\theta_s\omega) \in B_{\sqrt{\epsilon/(5a)}}(0)$. We derive from (5.36) that for such $\omega \in E$, we have

$$\|\Phi(T, \omega)\| \geq e^{T(\alpha-\frac{4}{5}\epsilon)} \geq K e^{(\gamma-\epsilon)T} .$$

This contradicts the assumption.

(ii) $\operatorname{rk} P_\gamma \equiv 1$. The argument is the same as in the previous case, since our estimates do not depend on the tangent vector v but hold for the norm $\|\Phi(t, \omega)\|$.

(iii) $P_\gamma \equiv 0$. This means that for almost all $\omega \in \Omega$, we have

$$\|\Phi(t, \omega)\| \geq \frac{1}{K} e^{(\gamma+\epsilon)t} \quad \text{for all } t \geq 0 .$$

Together with (5.35), this implies that

$$\frac{\ln K + (\alpha - \epsilon - \gamma)t}{a - |b|} \geq \int_0^t \|A(\theta_s\omega)\|^2 \mathrm{d}s .$$

Choose some $T > 1$ and $y \in \mathbb{R}^2$ such that

$$\|y\|^2 > 4 \max\left\{\frac{\ln K}{a - |b|}, \frac{\alpha - \epsilon - \gamma}{a - |b|}\right\} .$$

Take $\delta < \frac{\|y\|}{2}$. Then by Proposition 5.4, there exists a set $E \in \mathcal{F}_{-\infty}^T$ such that

$$A(\theta_s\omega) \in B_\delta(y) \quad \text{for all } s \in [0, T] \text{ and } \omega \in E .$$

This implies

$$\int_0^t \|A(\theta_s\omega)\|^2 \mathrm{d}s > T\frac{\|y\|^2}{4} > \frac{\ln K + (\alpha - \epsilon - \gamma)T}{a - |b|} ,$$

which is a contradiction.

This finishes the proof of this theorem.

We demonstrate now that the change of sign in the dichotomy spectrum is mirrored by finite-time properties of the system. To see this, consider a compact time interval $[0, T]$ and the corresponding *finite-time largest Lyapunov exponents* associated with the attracting random equilibrium $A : \Omega \to \mathbb{R}^2$, given by

$$\lambda^{T,\omega} := \sup_{\|v\|=1} \frac{1}{T} \ln \|\Phi(T,\omega)v\| \quad \text{for all } \omega \in \Omega\,.$$

From Proposition 5.3 (iii), we obviously have $\lambda_{\text{top}} = \lim_{T\to\infty} \lambda^{T,\omega}$ almost surely, where λ_{top} is the largest Lyapunov exponent of (5.23) as given in (5.31).

Proposition 5.5 *Consider the stochastic differential equation* (5.23) *with* $|b| < a$ *and* $\alpha \in \mathbb{R}$ *such that there exists a unique attracting random equilibrium* $A : \Omega \to \mathbb{R}^2$. *The following statements hold.*

(i) For $\alpha < 0$, *we have* $\lambda^{T,\omega} \le \alpha < 0$ *for all* $\omega \in \Omega$, *which means that the random attractor* $A : \Omega \to \mathbb{R}^2$ *is* finite-time attracting.
(ii) For $\alpha > 0$, *we have* $\mathbb{P}\big(\omega \in \Omega : \lambda^{T,\omega} > 0\big) > 0$, *which means that the random attractor* $A : \Omega \to \mathbb{R}^2$ *is not finite-time attracting.*

Proof

(i) Recall from (5.35) that

$$\|\Phi(t,\omega)\| \le \exp\left(\int_0^t \left(\alpha - (a - |b|)\|A(\theta_s\omega))\|^2\right) \mathrm{d}s\right),$$

which implies that

$$\lambda^{T,\omega} \le \frac{1}{T}\int_0^T \left(\alpha - (a - |b|)\|A(\theta_s\omega))\|^2\right) \mathrm{d}s \le \alpha < 0\,.$$

(ii) Recall from (5.36) that

$$\|\Phi(t,\omega)\| \ge \exp\left(\int_0^t \left(\alpha - 4a\|A(\theta_s\omega))\|^2\right) \mathrm{d}s\right).$$

Choose $\epsilon := \sqrt{\frac{\alpha}{5a}} > 0$. According to Proposition 5.4, there exists a set $E \in \mathcal{F}_{-\infty}^T$ of positive measure such that $A(\theta_s\omega) \in B_\epsilon(0)$ for all $s \in [0,T]$ and $\omega \in E$. Then

$$\lambda^{T,\omega} \ge \frac{1}{T}\int_0^T \left(\alpha - 4a\|A(\theta_s\omega))\|^2\right) \mathrm{d}s \ge \alpha - \frac{4\alpha}{5} = \frac{\alpha}{5} > 0 \quad \text{for all } \omega \in E\,.$$

This shows the claim.

As a combination of Theorem 5.10, Theorem 5.11 and Proposition 5.5, we arrive at the following theorem about bifurcation of (5.23) when the shear is small.

Theorem 5.12 (Bifurcation for Small Shear) *Consider the stochastic differential equation* (5.23) *with* $|b| < a$. *Then the random attractor* A *is given by an attracting* *random equilibrium for all* $\alpha \leq 0$ *and all* $\alpha > 0$ *in a neighbourhood of* 0*. We observe the following bifurcation at* $\alpha = 0$*:*

(i) For $\alpha < 0$*, the random equilibrium is globally uniformly attractive, but for* $\alpha > 0$*, the random equilibrium is not even locally uniformly attractive.*

(ii) Let $\Phi(t, \omega) := \mathrm{D}D\varphi(t, \omega, A(\omega))$ *denote the linearized random dynamical system along the random equilibrium for fixed* α*. Then the dichotomy spectrum* Σ *of* Φ *is given by*

$$\Sigma = [-\infty, \alpha] ,$$

i.e., hyperbolicity is lost at $\alpha = 0$*.*

(iii) For $\alpha < 0$*, the random equilibrium is finite-time attractive, whereas for* $\alpha > 0$*, it is not finite-time attractive.*

References

1. Adrianova, L.Y.: Introduction to Linear Systems of Differential Equations, Translations of Mathematical Monographs, vol. 146. American Mathematical Society, Providence (1995)
2. Ambrose, W.: Representation of ergodic flows. Ann. Math. **42**(3), 723–739 (1941)
3. Anderson, B.D.O., Ilchmann, A., Wirth, F.R.: Stabilizability of linear time-varying systems. Syst. Control Lett. **62**(9), 747–755 (2013)
4. Anh, P.T., Doan, T.S.: Explicit formulas for Lyapunov exponents of planar bilinear stochastic differential equations. Stoch. Analy. Appl. **35**(4), 662–676 (2017)
5. Arbieto, A., Bochi, J.: L^p-generic cocycles have one-point Lyapunov spectrum. Stoch. Dyn. **3**, 73–81 (2003). (Corrigendum: Stoch. Dyn. **3**, 419–420 (2003))
6. Arnold, L.: Random Dynamical Systems. Springer, Berlin (1998)
7. Arnold, L.: A formula connecting sample and moment stability of linear stochastic systems. SIAM J. Appl. Math. **44**(4), 793–802 (1984)
8. Arnold, L., Cong, N.D.: Linear cocycles with simple Lyapunov spectrum are dense in L^∞. Ergodic Theory Dyn. Syst. **19**(6), 1389–1404 (1999)
9. Arnold, L., Kliemann, W., Oeljeklaus, E.: Lyapunov exponents of linear stochastic systems. In: Lyapunov Exponents (1984). Lecture Notes in Mathematics, vol. 1186, pp. 85–125. Springer, Berlin (1986)
10. Arnold, L., Gundlach, V.M., Demetrius, L.: Evolutionary formalism for products of positive random matrices. Ann. Appl. Probab. **4**, 859–901 (1994)
11. Arnold, L., Bleckert, G., Schenk-Hoppé, K.R.: The stochastic Brusselator: parametric noise destroys Hopf bifurcation. In: Crauel, H., Gundlach, M. (Eds.) Stochastic Dynamics, pp. 71–92. Springer, Berlin (1999)
12. Aulbach, B., Siegmund, S.: The dichotomy spectrum for noninvertible systems of linear difference equations. J. Differ. Equ. Appl. **7**(6), 895–913 (2001)
13. Aulbach, B., Wanner, T.: Integral manifolds for Carathéodory type differential equations in Banach spaces. In: Six Lectures on Dynamical Systems, pp. 45–119. World Scientific, Singapore (1996)
14. Aulbach, B., Wanner, T.: Topological simplification of nonautonomous difference equations. J. Differ. Equ. Appl. **12**, 283–296 (2006)
15. Aulbach, B., Pötzsche, C., Siegmund, S.: A smoothness theorem for invariant fiber bundles. J. Dyn. Differ. Equ. **14**(3), 519–547 (2002)
16. Barabanov, E.A., Konyukh, A.V.: Uniform exponents of linear systems of differential equations. Differentsial'nye Uravneniya **30**(10), 1665–1676 (1994)
17. Barabanov, E.A., Konyukh, A.V.: Bohl exponents of linear differential systems. Mem. Differ. Equ. Math. Phys. **24**, 151–158 (2001)

T. S. Doan, *Spectral Theory of Nonautonomous Dynamical Systems and Applications*, https://doi.org/10.1007/978-981-97-5520-2

18. Barreira, L., Pesin, Y.B.: Lyapunov Exponents and Smooth Ergodic Theory, University Lecture Series, vol. 23. American Mathematical Society, Providence (2002)
19. Battelli, F., Palmer, K.J.: Criteria for exponential dichotomy for triangular systems. J. Math. Analy. Appl. **428**(1), 525–543 (2015)
20. Baxendale, P.H.: Statistical equilibrium and two-point motion for a stochastic flow of diffeomorphisms. In: Spatial Stochastic Processes, Progress in Probability, vol. 19, pp. 189–218. Birkhäuser Boston, Boston (1991)
21. Baxendale, P.H.: A stochastic Hopf bifurcation. Probab. Theory Related Fields **99**(4), 581–616 (1994)
22. Berger, T.: Bohl exponent for time-varying linear differential-algebraic equations. Int. J. Control **85**(10), 1433–1451 (2012)
23. Bierkens, J., van Gaans, O., Verduyn Lunel, S.: Estimate on the pathwise Lyapunov exponent of linear stochastic differential equations with constant coefficients. Stoch. Analy. Appl. **28**(5), 747–762 (2010)
24. Bochi, J.: Discontinuity of the Lyapunov exponent for non-hyperbolic cocycles. Unpublished (1999)
25. Bochi, J.: Genericity of zero Lyapunov exponents. Ergodic Theory Dyn. Syst. **22**(6), 1667–1696 (2002)
26. Bochi, J., Viana, M.: Uniform (projective) hyperbolicity or no hyperbolicity: A dichotomy for generic conservative maps. Annales de l'Institut Henri Poincaré C, Analyse non linéaire **19**, 113–123 (2002)
27. Bochi, J., Viana, M.: The Lyapunov exponents of generic volume preserving and sympletic systems. Ann. Math. **161**, 1423–1485 (2005)
28. Bohl, P.: Über Differentialungleichungen. J. für die Reine und Angewandte Mathematik **144**, 284–313 (1914)
29. Bonckaert, P.: On the continuous dependence of the smooth change of coordinates in parametrized normal form theorems. J. Differ. Equ. **106**, 107–120 (1993)
30. Bonckaert, P.: Partially hyperbolic fixed points with constraints. Trans. Am. Math. Soc. **348**(3), 997–1011 (1996)
31. Bonckaert, P.: Conjugacy of vector fields respecting additional properties. J. Dyn. Control Syst. **3**(3), 419–432 (1997)
32. Bonckaert, P., Dumortier, F.: On a linearization theorem of Sternberg for germs of diffeomorphisms. Mathematische Zeitschrift **185**(1), 115–135 (1984)
33. Bonckaert, P., De Maesschalck, P., Dumortier, F.: Well adapted normal linearization in singular perturbation problems. J. Dyn. Differ. Equ. **23**(1), 115–139 (2011)
34. Bonckaert, P., De Maesschalck, P., Doan, T.S., Siegmund, S.: Partial linearization for planar nonautonomous differential equations. J. Differ. Equ. **258**, 1618–1652 (2015)
35. Bruhat, F.: Travaux de Sternberg. (French) [Works of Sternberg]. Séminaire Bourbaki **6**, Exp. No. 217, 179–196 (1995)
36. Buckwar, E., Kelly, C.: Asymptotic and transient mean-square properties of stochastic systems arising in ecology, fluid dynamics, and system control. SIAM J. Appl. Math. **74**(2), 411–433 (2014)
37. Bylov, B.F., Vinograd, R.E., Grobman, D.M., Nemytskii, V.V.: Theory of Lyapunov Exponents. Nauka, Moscow (1966) (In Russian)
38. Callaway, M., Doan, T.S., Lamb, J., Rasmussen, M.: The dichotomy spectrum for random dynamical systems and pitchfork bifurcations with additive noise. Annales de l'Institut Henri Poincaré, Probabilités et Statistiques **53**, 1548–1574 (2017)
39. Chu, J., Liao, F.-F., Siegmund, S., Xia, Y., Zhang, W.: Nonuniform dichotomy spectrum and reducibility for nonautonomous equations. Bull. des Sci. Math. **139**(5), 538–557 (2015)
40. Cornfeld, I.P., Fomin, S.V., Sinaĭ, Ya.G.: Ergodic Theory Grundlehren der Mathematischen Wissenschaften, vol. 245. Springer, New York (1982)
41. Colonius, F., Kliemann, W.: The Dynamics of Control. Birkhäser, Basel (2000)
42. Cong, N.D.: A remark on non-uniform property of linear cocycles. Vietnam J. Math. **28**, 81–85 (2000)

43. Cong, N.D.: A generic bounded linear cocycle has simple Lyapunov spectrum. Ergodic Theory Dyn. Syst. **25**, 1775–1797 (2005)
44. Cong, N.D., Doan, T.S.: An open set of unbounded cocycles with simple Lyapunov spectrum and no exponential separation. Stoch. Dyn. **7**(3), 335–355 (2007)
45. Cong, N.D., Doan, T.S.: On intergral separation of bounded linear random differential equations. Discrete Contin. Dyn. Syst. Seri. S **9**, 995–1007 (2016)
46. Constantine, G.M., Savits, T.H.: A multivariate Faà di Bruno formula with applications. Trans. Am. Math. Soc. **348**(2), 503–520 (1996)
47. Coppel, W.A.: Dichotomies in Stability Theory. Lecture Notes in Mathematics, vol. 629. Springer, Berlin (1978)
48. Crauel, H.: Markov measures for random dynamical systems. Stoch. Stoch. Rep. **37**(3), 153–173 (1991)
49. Crauel, H.: Lyapunov exponents of random dynamical systems on Grassmannians. In: Lyapunov Exponents (Oberwolfach, 1990). Lecture Notes in Mathematics, vol. 1486, pp. 38–50. Springer, Berlin (1991)
50. Crauel, H., Flandoli, F.: Additive noise destroys a pitchfork bifurcation. J. Dyn. Differ. Equ. **10**(2), 259–274 (1998)
51. Crauel, H., Kloeden, P.E.: Nonautonomous and random attractors. Jahresbericht der Deutschen Mathematiker-Vereinigung **117**(3), 173–206 (2015)
52. Crauel, H., Imkeller, P., Steinkamp, M.: Bifurcation of one-dimensional stochastic differential equation. In: Crauel, H., Gundlach, M. (Eds.) Stochastic Dynamics, pp. 27–47. Springer, Berlin (1999)
53. Cuong, L.V., Doan, T.S., Siegmund, S.: A Sternberg theorem for nonautonomous differential equations. J. Dyn. Differ. Equ. **31**(4), 1279–1299 (2019)
54. Daleckiĭ, J.L., Kreĭn, M.G.: Stability of Solutions of Differential Equations in Banach Spaces, Translations of Mathematical Monographs, vol. 43. American Mathematical Society, Providence (1974)
55. DeVille, L., Sri Namachchivaya, N., Rapti, Z.: Stability of a stochastic two-dimensional non-Hamiltonian system. SIAM J. Appl. Math. **71**(4), 1458–1475 (2011)
56. Doan, T.S.: Analyticity of Lyapunov exponent for generic bounded linear random dynamical systems. Discrete Contin. Dyn. Syst. B **22**(8), 3113–3126 (2017)
57. Doan, T.S., Palmer, K., Rasmussen, M.: Bohl spectrum for nonautonomous differential equations. J. Dyn. Differ. Equ. **29**(4), 1459–1485 (2017)
58. Doan, T.S., Engel, M., Lamb, J.S.W., Rasmussen, M.: Hopf bifurcation with additive noise. Nonlinearity **31**, 4567–4601 (2018)
59. Duan, J., Kan, X., Schmalfuss, B.: Canonical sample spaces for random dynamical systems. In: Perspectives in Mathematical Sciences. Interdisciplinary Mathematical Sciences, vol. 9, pp. 53–70. World Scientific, Hackensack (2010)
60. Dubois, L.: Real cone contractions and analyticity properties of the characteristic exponents. Nonlinearity **21**, 2519–2536 (2008)
61. Dumortier, F., Rodrigues, P.R., Roussarie, R.: Germs of Diffeomorphisms in the Plane. Lecture Notes in Mathematics, vol. 902. Springer, Berlin (1981)
62. Fabbri, R., Johnson, R.: Genericity of exponential dichotomy for two-dimensional differential systems. In: Annali di Matematica Pura ed Applicata. Serie Quarta **178**, 175–193 (2000)
63. Flandoli, F., Gess, B., Scheutzow, M.: Synchronization by noise. Probab. Theory Related Fields **168**(3–4), 511–556 (2017)
64. Flandoli, F., Schmalfuß, B.: Random attractors for the 3-D stochastic Navier-Stokes equation with mulitiplicative white noise. Stoch. Stoch. Rep. **59**(1& 2), 21–45 (1996)
65. Froyland, G., González-Tokman, C., Quas, A.: Stochastic stability of Lyapunov exponents and Oseledets splitting for semi-invertible matrix cocycles. Commun. Pure Appl. Math. **68**, 2052–2081 (2015)
66. Garrido-Atienza, M.J., Schmalfuss, B.: Ergodicity of the infinite dimensional fractional Brownian motion. J. Dyn. Differ. Equ. **23**(3), 671–681 (2011)

67. Halmos, P.R.: Lectures on Ergodic Theory. Chelsea Publishing, New York (1960)
68. Haller, G.: Finding finite-time invariant manifolds in two-dimensional velocity fields. Chaos **10**(1), 99–108 (2000)
69. Haller, G., Yuan, G.: Lagrangian coherent structures and mixing in two-dimensional turbulence. Phys. D Nonlinear Phenom. **147**(3 & 4), 352–370 (2000)
70. Hardi, M.: Combinatorics of partial derivatives. Electr. J. Combinat. **13**, no. 1, Research Paper 1, 13 (2006)
71. Hartmanm, P.: Ordinary Differential Equations. Birkhäuser, Boston (1982)
72. Hinrichsen, D., Ilchmann, A., Pritchard, A.J.: Robustness of stability of time-varying linear systems. J. Differ. Equ. **82**(2), 219–250 (1989)
73. Ikeda, N., Watanabe, S.: Stochastic Differential Equations and Diffusion Processes. North Holland-Kodansha, Tokyo (1981)
74. Imkeller, P., Lederer, C.: An explicit description of the Lyapunov exponents of the noisy damped harmonic oscillator. Dyn. Stabil. Syst. **14**(4), 385–405 (1999)
75. Imkeller, P., Lederer, C.: Some formulas for Lyapunov exponents and rotation numbers in two dimensions and the stability of the harmonic oscillator and the inverted pendulum. Dyn. Syst. **16**(1), 29–61 (2001)
76. Imkeller, P., Lederer, C.: The cohomology of stochastic and random differential equations, and local linearization of stochastic flows. Stoch. Dyn. **2**(2), 131–159 (2002)
77. Imkeller, P., Schmalfuß, B.: The conjugacy of stochastic and random differential equations and the existence of global attractors. J. Dyn. Differ. Equ. **13**(2), 215–249 (2001)
78. Jan, L.: Equilibre statistique pour les produits de diffeomorphismes aleatoires independants. Annales de l'Institut Henri Poincaré, Probabilitées et Statistiques **23**(1), 111–120 (1987)
79. Kato, T.: Perturbation Theory for Linear Operators. Springer, Berlin (1980)
80. Keller, H., Ochs, G.: Numerical approximation of random attractors. In: Crauel, H., Gundlach, M. (Eds.) Stochastic Dynamics, pp. 93–115. Springer, Berlin (1999)
81. Keller, H., Schmalfuss, B.: Attractors for stochastic differential equations with nontrivial noise. Buletinul A.S. a R.M. Matematica **26**(1), 43–54 (1998)
82. Kloeden, P.E., Platen, E.: Numerical Solution of Stochastic Differential Equations, Applications of Mathematics (New York), vol. 23. Springer, Berlin (1992)
83. Kloeden, P.E., Rasmussen, M.: Nonautonomous Dynamical Systems, Mathematical Surveys and Monographs, vol. 176. American Mathematical Society, Providence (2011)
84. Knill, O.: The upper Lyapunov exponent of $Sl(2; \mathbb{R})$ cocycles: Discontinuity and the problem of positivity. In: Lyapunov Exponents. Lecture Notes in Mathematics, vol. 1486, pp. 86–97 (1990)
85. Li, W., Lu, K.: Poincaré theorems for random dynamical systems. Ergodic Theory Dyn. Syst. **25**, 1221–1236 (2005)
86. Li, W., Llibre, J., Wu, H.: Normal forms for almost periodic differential systems. Ergodic Theory Dyn. Syst. **29**(2), 637–656 (2009)
87. Liao, L., Li, M., Wu, H.: General laws of the analytic linearization for random diffeomorphisms. Mathematische Zeitschrift **270**, 739–757 (2012)
88. Linh, V.H., Mehrmann, V.: Lyapunov. Bohl and Sacker-Sell spectral intervals for differential-algebraic equations. J. Dyn. Differ. Equ. **21**(1), 153–194 (2009)
89. Lyapunov, A.M.: The General Problem of the Stability of Motion. Mathematical Society of Kharkov, Kharkov (1892) (In Russian)
90. Malkin, I.G.: A theorem on stability in the first approximation. Doklady Akad. Nauk SSSR (N.S.) **76**, 783–784 (1951)
91. Mierczyński, J., Shen, W.: Spectral theory for forward nonautonomous parabolic equations and applications, Fields Inst. Commun. **64**, 57–99 (2013)
92. Millionščikov, V.M.: Systems with integral separateness which are everywhere dense in the set of all linear systems of differential equations. Differ. Equ. **5**, 850–852 (1969)
93. Namachchivaya, N.S.: Stochastic bifurcation. Appl. Math. Comput. **38**(2), 101–159 (1990)
94. Neirynck, K.: Local equivalence and conjugacy of families of vector fields and diffeomorphisms. Diss UHasselt Diepenbeek (2005)

95. Nelson, E.: Topics in Dynamics I: Flows, Princeton University Press, Princeton (1969)
96. Oseledets, V.I.: A multiplicative ergodic theorem. Lyapunov characteristic numbers for dynamical systems. Trans. Moscow Math. Soc. **19**, 197–231 (1968)
97. Palmer, K.J.: A generalization of Hartman's linearization theorem. J. Math. Analy. Appl. **41**, 753–758 (1973)
98. Palmer, K.J.: A diagonal dominance criterion for exponential dichotomy. Bull. Austral. Math. Soc. **17**(3), 363–374 (1977)
99. Palmer, K.J.: A characterization of exponential dichotomy in terms of topological equivalence. J. Math. Analy. Appl. **69**(1), 8–16 (1979)
100. Palmer, K.J.: Exponential separation, exponential dichotomy and spectral theory for linear systems of ordinary differential equations. J. Differ. Equ. **46**, 324–345 (1982)
101. Palmer, K.J., Siegmund, S.: Generalized attractor-repeller pairs, diagonalizability and integral separation. Adv. Nonlinear Stud. **4**, 189–207 (2004)
102. Pötzsche, C.: Geometric Theory of Nonautonomous Discrete Dynamical Systems and Discretizations. Lecture Notes in Mathematics, vol. 1907. Springer, Berlin (2010)
103. Pötzsche, C., Rasmussen, M.: Computation of nonautonomous invariant and inertial manifolds. Numer. Math. **112**(3), 449–483 (2009)
104. Pötzsche, C., Siegmund, S.: C^m-smoothness of invariant fiber bundles. Topol. Methods Nonlinear Analy. **24**, 107–147 (2004)
105. Pollicott, M.: Maximal Lyapunov exponents for random matrix products. Invent. Math. **181**(1), 209–226 (2010)
106. Rasmussen, M.: Dichotomy spectra and Morse decompositions of linear nonautonomous differential equations. J. Differ. Equ. **246**(6), 2242–2263 (2009)
107. Ruelle, D.: Analyticity properties of the characteristic exponents of random matrix products. Adv. Math. **32**(1), 68–80 (1979)
108. Sacker, R.J., Sell, G.R.: A spectral theory for linear differential systems. J. Differ. Equ. **27**, 320–358 (1978)
109. Sell, G.R.: Topological Dynamics and Ordinary Differential Equations. Van Nostrand Reinhold Mathematical Studies, London (1971)
110. Siegmund, S.: Spektraltheorie, glatte Faserungen und Normalformen für Differentialgle ichungen vom Carathéodory-Typ, Dissertation University of Augsburg (1999)
111. Siegmund, S.: Normal form of Duffing-van der Pol oscillator under nonautonomous parametric perturbations. In: Proceedings of International Conference on Dynamical Systems and Differential Equations, Atlanta, Published as Added Volume of Discrete and Continuous Dynamical Systems, pp. 357–361. (2000)
112. Siegmund, S.: Dichotomy spectrum for nonautonomous differential equations. J. Dyn. Differ. Equ. **14**(1), 243–258 (2002)
113. Siegmund, S.: Reducibility of nonautonomous linear differential equations. J. Lon. Math. Soc. **65**(2), 397–410 (2002)
114. Siegmund, S.: Normal forms for nonautonomous differential equations. J. Differ. Equ. **178**(2), 541–573 (2002)
115. Siegmund, S.: Normal forms for nonautonomous difference equations. Comput. Math. Appl. **45**, 1059–1073 (2003)
116. Sternberg, S.: Local contractions and a theorem of Poincaré. Am. J. Math. **79**, 809–824 (1957)
117. Sternberg, S.: On the structure of local homeomorphisms of Euclidian n-space, I. Am. J. Math. **80**, 623–631 (1958)
118. Takens, F.: Partially hyperbolic fixed points. Topology **10**, 133–147 (1971)
119. Tearne, O.M.: Boundary approximation of deterministic and random attractors. Nonlinearity **18**(5), 2011–2034 (2005)
120. Tearne, O.M.: Collapse of attractors for ODEs under small random perturbations. Probab. Theory Related Fields **141**(1–2), 1–18 (2008)
121. Tong, H.: Some characterizations of normal and perfectly normal spaces. Duke Math. J. **19**, 289–292 (1952)

122. Walter, P.: An Introduction to Ergodic Theory, Graduate Texts in Mathematics, vol. 79. Springer, New York (1982)
123. Wang, B.: Stochastic bifurcation of pathwise random almost periodic and almost automorphic solutions for random dynamical systems. Discrete Contin. Dyn. Syst. Ser. A **35**, 3745–3769 (2015)
124. Wirth, F.: Dynamics of time-varying discrete-time linear systems: spectral theory and the projected system. SIAM J. Control Optimiz. **36**(2), 447–487 (1998)
125. Whittlesey, E.F.: Analytic functions in Banach spaces. Proc. Amer. Math. Soc. **16**, 1077–1083 (1965)
126. Wu, H., Li, W.: Poincaré type theorems for non-autonomous systems. J. Differ. Equ. **245**(10), 2958–2978 (2008)

Index

T. S. Doan, *Spectral Theory of Nonautonomous Dynamical Systems and Applications*, https://doi.org/10.1007/978-981-97-5520-2

P

R

S

T

U

W

FSC
www.fsc.org
MIX
Papier aus verantwortungsvollen Quellen
Paper from responsible sources
FSC® C105338

If you have any concerns about our products,
you can contact us on
ProductSafety@springernature.com

In case Publisher is established outside the EU,
the EU authorized representative is:
Springer Nature Customer Service Center GmbH
Europaplatz 3, 69115 Heidelberg, Germany

Printed by Libri Plureos GmbH
in Hamburg, Germany